The Autobiography of Light

George Grammatikakis

The Autobiography of Light

 Springer

George Grammatikakis
Heraklion, Greece

Translated by
Ben Petre
Freelance Translator
Heraklion, Greece

ISBN 978-3-031-56916-6   ISBN 978-3-031-56917-3   (eBook)
https://doi.org/10.1007/978-3-031-56917-3

Translation from the Greek language edition: "The Autobiography of Light" by Ben Petre, © Crete University Press 2005. Published by Crete University Press. All Rights Reserved.

Cover photograph: Danae Syrrou, "An afternoon room reflection" (photography, 2019)

This Springer imprint is published by the registered company Springer Nature Switzerland AG
The registered company address is: Gewerbestrasse 11, 6330 Cham, Switzerland

If disposing of this product, please recycle the paper.

Preface

The Autobiography of Light was originally published in November 2005 by Crete University Press, one of the most reputable publishers in Greece. The first edition aroused lively interest and won critical acclaim, enjoying a long stay on the best-seller lists. That in turn led to the book being constantly reprinted, though its content and length remained virtually unchanged.

Yet one of the main features of our time, which in other respects holds many ills in store for humans and their lives, is that science continues to chart an impressive, almost unimpeded forward course. Physics and cosmology in particular, which are the general topics of this book, have made astonishing achievements and progress over the past two decades. Consequently, the prospect of publishing an English-language edition led to the need to enhance the content to make it fully reflect the present state of science.

The Autobiography of Light thus reaches out to a global audience enriched with all the latest developments in science, while standing every bit as proud of its record to date, accompanied by praise and well-loved by its readers. At all events, the book's new journey would not have been possible without the knowledge, diligence and hard work of Ben Petre, who translated *The Autobiography* into English and edited the bibliography. My colleague and friend Stefanos Trachanas was on hand, ever eager to discuss any difficulties large or small that arose. Invaluable input was provided by yet another colleague and friend, Vassilis Charmandaris, who meticulously combed through the English text and made apt and insightful observations. Thanks are also due to my friend Pantelis Ikonomou, for useful conversations. It goes without saying that the biggest debt is to my family, for their unfailing support.

While this new edition stands out for its refreshed, updated form, it also retains the core that won it recognition. And that is the staggering realisation that light and its wondrous pathways have led science to its greatest accomplishments: to the theory of relativity, the quantum behaviour of the microcosm, and to modern views on the birth and evolution of the universe. At the same time, the book underscores the importance of light for life itself and for art, while also referring to the philosophical or religious views that have accompanied light down the course of centuries.

In sum, *The Autobiography of Light* uses simple language to describe the fascinating adventure of modern science, which has light as its unerring, revelatory guide. And that is why in the book's last chapter, the floor is metaphorically taken by primordial light itself, one of the greatest discoveries of modern science. In its own words, that light narrates its own aeons-long course through the universe, and its revelatory encounter with humankind. My hope is that my readers will be just as charmed as I have been by the infinite pathways of light, and that they will savour the pages of this book in their minds, and either in parallel or at some later stage in their hearts.

Heraklion, Greece   George Grammatikakis

Prologue: The Author and Light

He told me his book
was called the Book of Sand
because neither the book nor sand
has any beginning or end

—*Jorge Luis Borges*

Out of honesty towards readers, it ought to be confessed from the outset that the autobiography of light is, deep down, a love story. At some moment lost in time, light emerged unprompted at the centre of the author's life. Its colours and enigmatic nature, the infinity of its pathways and its iridescences seemed to possess an ineffable charm. So they were not long in conquering his soul. Just like that woman, if she existed—what pathway will she now be travelling on, I wonder—who appeared opposite him at some point, on the deck of a ship, and silently followed him from island to island and from sea to sea, for years and for centuries.

The truth is that I, the author, cannot pinpoint exactly when this adventure with light began, though I do remember that as a child I was afraid of nighttime. I could feel the night enveloping and threatening me, so I'd often stay awake waiting for the light. In the morning, redeeming light would come creeping in through crevices and chasing the night away, brightening as time passed and giving things identity. It was a calm, tender light, yet at other times—in summers—it would soon grow dazzling and merciless.

This game with light and the fear of nighttime went on over many years. It's still with me now, perhaps less acute. I usually only sleep a few hours, and even in the night I get up and wander around aimlessly. It's as if the arrival of light—like *her* arrival—pours balsam on the troubled ways of the soul.

I already knew a few things about light from school, though in a rather piecemeal way and without realising their importance: the laws of reflection and refraction, interference phenomena and the incomprehensible polarisation of light. But more tedium than all the rest was inflicted by an entire chapter of our *Optics* on lenses: convex and concave lenses, focal lengths, real or illusory images. Nearing the end of school, I believed that like other things in life, the light described in books had nothing to do with the light around us. There was just one day when a teacher spoke to us about photosynthesis. And the passion in his voice really did convince us that light had some important role to play in life itself.

Later, partly the trend for "sciences" and partly some fateful coincidences led me to the Physics Department at the University of Athens. There I tried to tease out some of the scientific concepts connected to light. Things definitely improved compared with school, yet rarely went beyond formalism and the surface. The vague expectation that university meant a distinct living and learning environment was rapidly confounded. Consequently, as is true of present-day students, my efforts focused on the demands made by exams. The rest of the time I preferred heading off to the cinema or theatre or spending hours browsing in bookshops.

In any case, I had come to understand that light was of fundamental significance to physics itself. To my surprise, I even realised that its nature was no different from that of radio waves or the X-rays used in medicine. Besides, the concept of "wave-particle duality", which seemed crucial, posed a profound conundrum to us students. We couldn't easily get our heads around either that concept or any other elements of quantum physics then taught at university. As was true of so many other things, we had become trapped in the classical ideas of science. The shallow teaching methods and usually off-putting university textbooks were no help in overcoming them. As a result, my physics degree was accompanied by the bitter sense that I had barely tasted the quality of scientific knowledge.

On the other hand, my life often headed off along diverging but meaningful paths. The lustre of Left-wing ideas led me to the realms of poetry, cinema and artistic pursuits. All of a sudden, I discovered that the concept of light maintained a stable, powerful presence there too, loaded with all kinds of symbolism. One line by George Seferis—whose meaning I'm still not certain of—made a profound impression on me:

Years ago you said: 'Deep down I am a matter of light.'
Yet in the poetry of Elytis, who had yet to gain his present renown, light emerged with force and continual doublings-back.

From some point on, then, my relationship with light kept taking two parallel paths. The first, that of science, exacting and yet enchanting, led to a constant deepening of knowledge, but was fraught with major difficulties. The second road led through art, poetry or philosophy, trying to probe hidden meanings and significations. Of course, it didn't take me long to see that while the two roads were parallel, they did intersect and had common points. Besides, the general theory of relativity had shown that the true geometry of the universe was non-Euclidean and that parallel straight lines did intersect.

It is perhaps superfluous to stress that the road of science called for method and strenuous effort. Finding this out was a torment when I began postgraduate studies at Imperial College London. I discovered that knowledge of modern physics—of relativity, quantum mechanics and elementary particles—was necessary for the nature of light to emerge with any clarity. Thus, the wave-particle duality characterising light was a more general and inviolable principle of quantum mechanics. By contrast, the laws of reflection and refraction that had, among other things, poisoned our schooldays, simply expressed one of light's key capabilities: in moving from one point to another, it always chose the route that took the least time!

Back then the thing at the forefront of physics research was one of the subject's age-old problems: the structure of matter and its fundamental particles. More out of intuition than knowledge, that was the area I chose to do my doctorate in. It was already obvious that the protons and neutrons making up the nucleus of an atom were not the most elementary constituents of matter. An entire world of particles was emerging from experiments and theoretical research. In successive transformations or combinations, they participated in the everlasting processes of the universe. Lone and proud among them stood the photon, i.e. the fundamental unit of light.

In this world of scientific research, seemingly without end or beginning, I have spent a great part of my life. I worked as a researcher in England and, having returned to Greece—it was 1973—at the Demokritos Institute of Nuclear and Particle Physics in Athens. But being complex and costly, necessary experiments were always carried out at the European Organisation for Nuclear Research in Geneva, also known as CERN. That is the site of large accelerators—even more powerful ones are now under construction—where elementary particles are accelerated to enormous speeds and then left to collide with each other. Like a child smashing a toy with a hammer, these collisions can reveal the internal structure of matter, and shed light on the interactions ruling its elementary level.

Experimental needs meant I had to make plenty of trips and often spend time in Geneva. Later on, I was even invited to work there for an extended period. The need to understand the nature of light as much as possible alleviated homesickness, as did the intensity of research work. In fact, I enjoyed observing the constant transformations of photons into electrons detected by special sensors. It may really seem incredible, but light is converted into matter!

As my knowledge increased, the scientific side of light gained a more distinct form in my mind: light was an electromagnetic wave propagated at colossal speed, yet was simultaneously particulate in nature. The photon, the fundamental particle of light, had a place all of its own among the matter and energy particles making up the world. A neat, detailed theory known as quantum electrodynamics had already been formulated, describing the interactions between photons and matter. According to it, the photon is a kind of messenger exchanged between matter particles, creating electromagnetic forces. In fact, the idea of exchange was later extended to interpret the remaining forces. In science just as elsewhere, light has always shown the way!

Anyway, for as many years as I was forced to live far from Greece, the longing for Greek light was torture. Besides, there was also the longing for the one—what pathways of light will she be travelling on now?—who accompanied my thought and dreams. This double longing led to constant wavering in my feelings. The expectation of returning often mingled with fears over what meaning it had.

Here a break is needed, as it's the moment in around 1980 when life, and thus my preoccupation with light, reached a major turning point. The University of Crete had just begun operation, and I was called on to participate in organising it and teaching in the Physics Department. A short while later I was elected professor and left the Demokritos Institute for good. As for my scientific research work, I now had few opportunities to work on the CERN experiments and large accelerators. All the same, one of the most paradoxical and interesting "fundamental" particles had begun to preoccupy the Greek scientific community: the neutrino. Its mass was infinitesimal, its charge—like that of the photon—was zero and, being incredibly penetrative, it could pass through the entire Earth without difficulty. A grand and difficult experiment that would become known by the acronym NESTOR (Neutrino Extended Submarine Telescope with Oceanographic Research) aimed to detect ghostly neutrinos in the depths of the sea at Pylos, as they arrived there from distant galaxies or stellar explosions. I began collaborating with the experiment, which was inspired and coordinated by Prof. Leonidas Resvanis at the University of Athens. At the same time, our attempt to detect the neutrino in the depths of the sea, where cosmic radiation could not reach, had a transcendental dimension to it that mitigated the experiment's difficulties.

There is of course no need to stress that in this new university-focused cycle in my life, too, I felt the need to understand light as much as possible. Indeed, I now had some unexpected assistance, which was simultaneously a torment: my students. When I was teaching electromagnetism or the theory of relativity—which are of fundamental importance to the theory of light— the questions they asked often put me on the spot. More than a few times I made a diplomatic sidestep. All the same, the experience of teaching rein-forced the suspicion from my student years that light and its behaviour were the cornerstones of physics and its major accomplishments. It's characteristic that the receding of the galaxies was revealed thanks to their light; and that the discovery of "primordial" light confirmed that the universe was created by a Big Bang. As well as the evolution of life on Earth, this impressive flowering of cosmology was described in my first book, *Berenice's Hair*, which proved surprisingly popular.

Light, then, proved to be an invaluable messenger of the universe's secrets. So the irresistible attraction I had been feeling for years filled me with pride; but it also contained traces of jealousy, which even then seemed inexcusable.

As for the other side of light, to do with art or its games with people and landscapes, it goes without saying that I let down my defences to an allure that knew no bounds. So it was no exaggeration when, at the begin-ning of this prologue (or perhaps confession?) I wrote that this book about light is, deep down, a love story. Science, meaning my rational mind, was constantly striving to understand light, to analyse its behaviour via laws and equations. Yet the presence of light itself—as with every great love—was bidding for unconditional acceptance, a surrender to passion and the allure of the uninterpretable. Perhaps that's why that woman, who had accompanied my wanderings for centuries, remained silent in the face of my insistence on guessing her intentions. "But," wrote Kafka, "the Sirens have an even more terrible weapon than song: their silence."

In any event, I discovered the true significance of light in our life and feelings when, as I have already related, the migrations of the soul led me to Greece for good. In England and the European countries I had lived in up until then, the light seemed as if it were gasping or being persecuted by some unknown, invisible enemy. But on my return to Greece, I got the feeling I had rediscovered light. "Greek light," wrote Heny Moore, "is, as everyone says, something that you cannot imagine unless you experience it. In England, half the light, in some way, is absorbed by objects. However, in Greece, objects seem to give off light, as though illuminating themselves from within." I genuinely began to be convinced that, as our teachers and poets have long stressed, light is entwined with Greece and its destiny. And even more so that I bore the expectation and need of light within me.

In that way, travels in my own country now took on a different, rather paradoxical content: I felt that I was reliving Greece as multiple quests for light, thanks too to the varied, revelatory forms it assumed. Really. In the islands of the Aegean and in Epirus, in Attica or the Ionian Sea, Greek light always laid its own peculiarities bare, lending a sense of identity to the area's inhabitants and geography. In fact, I was not long in expanding my findings. Here in Greece, the clarity of light has determined the character not only of the landscape, but also of our intellectual quests. An imaginary line starting from the Apollonian light of antiquity ran through the phrase "Come receive the light" in the Orthodox liturgy of the Resurrection, and reached down to Odysseas Elytis' "Worthy is the Light". As the poet himself notes, "Light and history in Greece are one and the same thing… the one replicates the other, the one interprets and vindicates the other, even the very void that is black."

So it is, then: in Greece, light does not simply relay images and faces, it is not merely the messenger of the universe. It is elevated to the level of creator, moulding and highlighting in some magic way.

In 1990, my election as chancellor at the University of Crete radically altered my way of life and everything in it. In the intensity and loneliness of the position, the light was always there as anguish and solace, meaning and refutation. That phrase by Seferis, "deep down I am a matter of light", which had hounded me since my student years, was now gaining a clearer content.

The kindness of my colleagues and students led me to a second term of office as chancellor. When it came to an end, *Cosmic Writings* came out. Like *Berenice's Hair*, this second and more personal book was much loved by readers. But I had already developed a tantalising desire to write a book about light. The idea didn't move along much, though, as there was no lack of new challenges. At around the same time, the Greek Broadcasting Corporation decided to make a television series out of *Berenice's Hair*; the filming took time and a lot of travelling which, as I thought at least, was distracting me from the idea behind the book. That turned out not to be the case. Light continued to accompany my quests and my thoughts. I remember that once we had to spend the night with the television crew on Delos. At first, we witnessed the games played by light at dusk, and then the gentle light of nighttime spread everywhere. The only thing accompanying the sleepless wait for morning was the light of the stars and the sounds of waves dying on the shore. It was as if time had stood still. Like then, like always, when the expectation of her cancelled out any other thought or intention. When at some point the evocative morning light of Delos broke, inscrutable and ineffable, I simply abandoned myself to its allure. The dazzling sun soon rose, burning our eyes and the landscape. Everything I had heard about the miracles of light on the sacred isle paled in the face of reality—and what is reality, one wonders.

From very far back, light has been a symbol encapsulating the good in the world. To the Persians, for instance, Ahura was the god of light and represented beauty, wisdom and kindness. Conversely, Ahriman was the lord of darkness and the personification of evil. According to Egyptian mythology, the Sun in the form of the god Ra was the creator of the world, who touched Earth and gave birth to its elements. Of those elements, fire and water played a leading role. The stars were believed to be the souls of dead kings who ascended into heaven to be united with the Sun god or join his retinue.

Later on, in the more refined Greek world, sun worship was associated with Apollo. As an entity, light has a substantial presence in theories of the universe's origins. According to Hesiod's *Theogony*—the earliest depiction of the Cosmos, and one of the finest—the leading trio of creation consisted of Chaos, Earth and Love. Night and Darkness were born of Chaos, and from their union came the opposites, Day and Aether. And so is expressed the standard pre-scientific knowledge that night is succeeded by day, and darkness gives way to light.

So time went by. The book writing was progressing haphazardly. First one thing would crop up, then another. The passing of time was in any case a devious force, always putting me under pressure. In despair, I would count the things I wanted to finish, while there were others I would have liked to start from scratch. But I was afraid there wasn't much time left.

This curious relationship with time may explain the interest in old pocket watches that I, the author, gained at some stage. Each watch has its own unique beauty, which seems untouched by the passage of time. Besides, while its lifespan often runs into centuries, the wondrous mechanism doesn't show substantial signs of wear either. Yet how many times might it have changed hands; how many life stories might it have witnessed! As I carefully wound the watches and hung them here and there with their wonderful chains, I had the sense that time did not exist. That it was breaking down into dozens of faces.

All the same, real time continued to flow relentlessly on. A new challenge—to take on the Ionian University as president of its Governing Board—made the prospect of the book recede again. On the other hand, it would have been difficult to decline this fruitful meeting with Ionian light. Sure enough, Corfu and the islands of the Ionian Sea had their own phantasmagorias of light to show off; and I would simply let down my defences to its allure. At the other end of the same geographic arc, at Pylos, the NESTOR experiment was continuing its titanic efforts to detect the neutrino. To that end, a sensitive electronic device resembling an outsized starfish had been submerged for trial purposes in the depths of the sea. As neutrino detection would—like light—yield important information on the stars and their evolution, and perhaps also on the birth of the universe itself, that major prospect buoyed up all of us taking part. Yet the demands and difficulties of the experiment often seemed intractable, and it looked as if a stop to it was not far off.

Though it may no longer seem so anymore, the paradoxical thing was that while time and commitments never ceased to hound me, writing the book kept on gaining momentum. It was solidifying within me both as a wish and as a kind of duty. As I believed at the outset, the reasons for that paradoxical need had a rationalism to them, in a good sense. I kept thinking that light was an all-important presence in our lives, so the scientific dimension to it would be of great interest to readers.

Besides, in science itself, light really does have a place of its own. It is difficult to find areas of knowledge untouched by its multifaceted presence. For physics in particular, to investigate light is to depict the history of the subject itself. A history that doesn't head in a straight line, but rather in a broken one: achievements go hand in hand with dissent, accomplishments with failures.

At any rate, it's worth underlining that just like many other things in life and in the mind, the concept of light seems deceptively familiar and simple. I hope you will be disabused of this delusion from the very first pages of this book. Like that woman who kept accompanying me on neverending wonderings, light persistently hides its true face, troubling our mind and soul with its secrets. Readers, careful readers, should perhaps ask themselves whether the same might not apply to other supposedly self-evident truths, and whether it might not be time to question the ease with which we arrive at assessments and conclusions. If the constant endeavour to understand light teaches anything, it is how often certainties dissolve, and how all kinds of dogmatism grow dangerous. I only hope that a sense of constant searching for an uncertain truth can touch readers with its beneficial power. They will then be less exposed to the easy and the ephemeral, to superficial judgements or the confusion over values that are symptoms of today's enfeebled culture.

So there were powerful objective reasons why it was worth writing a book about light. Besides, in *Berenice's Hair* I had noted that "a book is always an attempt to communicate. Whether the author is attempting to communicate with readers or themselves may not particularly matter. What does matter is when the communication leaves some internal traces on the author or the reader." It is obvious, I think, that the wondrous adventure of light has left internal traces on me, the author; it's those feelings I'm hoping to share with readers. But the time has come to confess that those traces have not only had to do with the importance of light or scientific understanding of it. As I have come to realise over time, there is also another substantial reason: it's that the nature of light is hallmarked by certain preposterous attributes, certain unique graces.

I want to be honest. From the time when the unexpected secrets hidden by light were revealed to my mind and soul, the attraction I felt grew immeasurable. That explains the fact that between the eddies of my life, I would always try to snatch some time, to move one chapter on, to correct another one. And that even in the small hours of the night, when the tidings of time and fear arrive, I would find myself wrestling with words, putting down on paper what readers will come across in the pages to follow.

Besides, we said from the outset that deep down, the autobiography of light is a love story. And like every such story, in its depths, it hides moments of serenity or passion, confirmation or refutation, grandeur or despair. In speaking of light, one speaks of the world and its secrets. Likewise—and this is perhaps more serious—one touches on some anxieties of the soul and its hidden questions.

Yet just as is true of any great love, my attraction towards light was accompanied by an unspoken jealousy. Or, to be more precise, there was a kind of admiration that also hid elements of jealousy within it. In any case, doesn't jealously go hand in hand with love, often clouding our reason and our feelings?

In its pages, then, this book will seek out the nature of light, describing its behaviour as far as possible. That way some of its important gifts will emerge. These gained particular importance the more I, the author—and perhaps also the reader—learned about myself. So my jealousy of light was at least grounded in something, it was less reprehensible or base.

Besides, the way the book got its title seems far from random. At some point, that woman who had been following my wanderings for centuries, from island to island and from sea to sea, drew closer in the faint starlight and came up to me. This time her presence was no illusion. I clearly saw the melancholy in her eyes and felt the touch of her hands on my own face. "Light," she said, "comes from very far off. From a Space where there was no space, and from a Time when there was no time. Ever since then it has enveloped the Cosmos and highlighted its every expression." She looked at me tenderly. "So the autobiography of light," she went on, "is my autobiography too. And it will never cease to be written for as long as I exist. As long as we exist," she whispered, and began to draw away again. In my confusion and passion, I didn't know whether it was her speaking or light itself.

That's how it happened, at any rate, and how the book got its title: *The Autobiography of Light*.

Contents

1 Light Gambols Through History   1
   1.1   The Illusion of Familiarity   1
   1.2   Flashes of Light in Antiquity   4
   1.3   The Platonic View of the World   8
   1.4   The Brilliant Ascendancy of Aristotle   11
   1.5   Geometric Light   14
   1.6   Light in the Thousand and One Nights   18
   1.7   Bursts of Light in the Middle Ages   21
   1.8   An Astronomer on Supplements   24
   1.9   A Messenger from the Stars   27
   1.10 The Rational Reconstruction of Knowledge   30
   1.11 The Luminous Parallels Converge   34

2 Light's Virtues and Vices   37
   2.1   A Scientific Prelude   37
   2.2   Light's Manifest and Hidden Graces   40
   2.3   The Wonderful World of Colours   42
   2.4   Chromatic and Philosophical Acrobatics   45
   2.5   A Speed Different from Others   48
   2.6   Isaac Newton's Bright Ideas   51
   2.7   Light Contributes to an Understanding of Itself   55
   2.8   Aether Waves and Delusions   58
   2.9   The Importance of Being Wrong   60
   2.10 The Fields of an Autodidact   62
   2.11 A Bolt from the Blue   64
   2.12 The Dream Begins and Ends   66

3   Light's Revelation of Time and Space   71
   3.1   Einstein's Magic Carpet   71
   3.2   A Lone Traveller Through the Universe   73
   3.3   Simple Axioms with Major Consequences   76
   3.4   Brilliant Answers to Old Questions   79
   3.5   The Merging of Time and Space   81
   3.6   The Age of Light   84
   3.7   Time Machines and Time-Makers   86
   3.8   The Perilous Journeys of Length and Mass   89
   3.9   A Paradoxical World Is Confirmed   92
   3.10   The Equation Typical of Our Times   94
   3.11   God Was a Geometer   97
   3.12   The Bible of the Universe   100
   3.13   The Discovery of the Century   103

4   The Quantum Realm of the Microcosm   107
   4.1   The Road to Quantum Theory   107
   4.2   Photons Are Here to Stay   110
   4.3   Quantum Leaps in the Atom   112
   4.4   The Dual Nature of Things   117
   4.5   God Does Play Dice with the World   121
   4.6   The Certainty of Ignorance   124
   4.7   Quantum Acrobatics   128
   4.8   The Photon Love Bond   131

5   The Lustrous Dream of Unification   135
   5.1   An Age-Old Quest   135
   5.2   The Unanticipated Road to Antimatter   138
   5.3   An Acronym Full of Meaning   141
   5.4   The Choreographer of Light   143
   5.5   Forces and Weaknesses   146
   5.6   On Along the Trail Blazed by Light   150
   5.7   Heavy Light   152
   5.8   Quarks and Their Eternal Bondage   156
   5.9   An Unfinished Theory of Everything—And Its Triumph   159
   5.10   Two Parallels That Converge   161

6   Light Shakes Hands with Life and Art   165
6.1   Light, the Driving Force of Life   165
6.2   The Incredible Miracle of Sight   168
6.3   And Man Said: "Let There Be Light!"   171
6.4   Humans Tame Light   174
6.5   Light in Creative Captivity   179
6.6   Light Collaborates in the Microcosm   182
6.7   Light That Burns and Light That Creates   185
6.8   Light's Wanderings in Art   187

7   The Universe: The Empire of Light   193
7.1   Light Betrays the Secret of the Galaxies   193
7.2   The Big Bang: A Day Without Yesterday   198
7.3   A Fossil of Light   201
7.4   Early Traces in the Formation of the Galaxies   205
7.5   The Language of the Stars   206
7.6   The Past Surrounds Us   210
7.7   Nuclear Reactors in the Stars   216
7.8   The Life and Times of the Stars   219
7.9   The Stellar Apotheosis of Light   222
7.10   Large Stars and Their Dark Prospects   226
7.11   Photographing the Invisible   228
7.12   A Dark Universe in Dark Times   233
7.13   A Dark Sea of Energy   235
7.14   Our Universal Prospects   238
7.15   Mankind's Whimper and the End of Light   240

8   Epilogue: A Farewell to Light   245

9   The Autobiography of Light   251

Notes   273

Index   285

About the Author

George Grammatikakis (1939–2023) Born in Heraklion, Crete, George Grammatikakis read Physics at the University of Athens before moving on to postgraduate work and research at Imperial College London, where he received his PhD in 1973. On returning to Greece he joined the staff at the Demokritos National Centre for Scientific Research, later working at the European Organisation for Nuclear Research (CERN) in Geneva. He had a key role to play in organising the newly founded University of Crete, where he was appointed professor of Physics, and subsequently professor emeritus. He was elected university chancellor for two terms (1990–1996), later going on to serve for four years as Chairman of the Governing Board at the University of the Ionian. Additionally, he participated in European Union international expert committees on education and research prospects.

His research interests centred on the structure of matter and cosmology, linking the microcosm to the macrocosm. As a visiting professor at Harvard University (1989–1990), he also worked on the history of science.

A versatile personality, he was a member of the Governing Board of the Greek Broadcasting Corporation, president of the Nikos Kazantzakis Museum on Crete and vice-president of the Greek National Opera. His belief both in bridging art and science and the importance of popularising science found a range of outlets. These ranged from press articles to artistic events featuring science narratives, images and music (including acclaimed performances in Athens and Thessaloniki based on *The Autobiography of Light*). As well as editing and producing successful public television documentaries, he created an award-winning series based on his book *Berenice's Hair*. The interrelationship between science, art and all manifestations of human culture permeated everything in his enduringly popular work, leading to multiple reprintings of his five books.

As an actively involved citizen with a genuine interest in people and public life, he was elected MEP in the 2014 European Parliament election, serving as a member of the Parliamentary Committee for Culture and Education, as well as the Committee for the Environment.

George Grammatikakis won numerous accolades and prizes in Greece, and was made a Knight of the Order of Academic Palms by the French Republic.

1

Light Gambols Through History

1.1 The Illusion of Familiarity

Light fills the entire universe. There is light at the most distant points in space, in the deep silence of night, in the faces and landscapes of our world. Yet its presence is only indirectly perceptible. We cannot see light itself. Via light we can see objects, their colours or how they move, as our brain synthesizes them. It is as if nature wished to protect us: if light itself were perceptible the way it arrives every moment and from all directions, it would overload our brain functions. The suspicion that something comes between us and the surrounding world arises from light's often paradoxical behaviour, like the tricks played by sunlight in the atmosphere, or an oar that appears to bend in seawater.

On the other hand, we sense that light is very familiar, making up the basic tissue of our life. In some sense we live in light, create under its wing, owe our existence to it. The relationship between humans and light is as essential as that between fish and water. Without light, neither human beings nor their life is conceivable. Apart from those scarred by misfortune, humans only spend moments of their life acting in the absence of light, devoid of its beneficial contribution. Besides, as we learnt in our schooldays, the wondrous process of photosynthesis is what maintains the cycle of life.

So it is that darkness—the absence of light—is quite rightly linked to fears, whether grounded or ungrounded, to death itself, and to the sinister powers in our own selves and in nature. On the other hand, thanks precisely to its special meaning, the word 'light' is loaded with symbolism. "Light of my life" is what lovers call each other; we Greeks never tire of repeating that Ancient

G. Grammatikakis, *The Autobiography of Light*, https://doi.org/10.1007/978-3-031-56917-3_1

Greek civilisation was "a radiant beacon" for humanity; and in celebrating the resurrection on Easter night, the Orthodox faithful chant: "Come receive the light, from the light that is never overtaken by night".

Down the course of the centuries, light has always been linked to wisdom and truth, whereas darkness represents Evil. And while this dichotomy is typical of the high points in the mythology and culture of every era, eighteenth century Europe was dominated by a broad intellectual and philosophical movement that was, in characteristic fashion, to be named the Enlightenment. The Enlightenment expressed faith in progress and the potential for humans to attain perfection, defended political freedoms and tolerance of others' convictions, and reacted against the irrational interpretation of the world and its submission to religious dogmas. Enlightenment ideas gained powerful political and social support, thus exercising considerable influence—something which, sadly, they still need to do!

For human beings, then, light is a basic precondition that enables them to perceive space and motion; it also enables them to perceive the colours and shape of objects. Light reveals the world, yet is also its creator.

It is true that the human eye often has the illusion that the sky is bright of its own accord, and that the brightness of the Earth and material bodies is a quality inherent in them, which is only lost in the dark. In reality, however, light travels in a dark universe before ending up on Earth and illuminating its sky. It is worth noting that almost all the information concerning the innumerable galaxies and stars reaches us in the form of light.

Besides, on a deeper level our world is made up of light and matter. The perennial interaction and interplay between the two often lie behind the world of phenomena. Indeed, as modern physics has revealed, light itself can be transformed into matter. Incredible as it may sound, this transformation is a run-of-the-mill experience in large research laboratories. What is more, modern cosmology maintains that the orgiastic transformation of light into matter particles—and vice-versa—dominated the first moments of creation.

Our life is not only bathed in natural light, which derives mainly from the sun. With fire as their distant, invaluable ancestor, various artificial light sources are now of vital importance to people and their activities. The production of electric light was obviously one important step that radically changed our social and personal habits. And in the laser, modern life is feeling the ever-increasing impact of another source of artificial light. As concentrated, high-intensity light, laser beams are already working wonders in everyday life, medicine and industry.

Yet it is not simply the physical presence of light nor its many useful forms that renders it our inseparable companion. It is also a central reference point in art. With the aid of light, the visual arts reproduce shapes, masses and forms. Since ancient times, the evocative power of colour has set its seal on masterpieces on vases and frescoes. Besides, two modern art forms—photography and the cinema—are founded on recording the games light plays on the people and things in our world.

The everlasting presence of light in our world and life leads to the grand illusion that nature itself is simple, and that its attributes and deeper structure are easy to grasp. Thus, down the course of the centuries, philosophers and major thinkers from Leucippus to Goethe and from Plato to Leonardo da Vinci formulated views on light and the operation of vision. Though they often showed intellectual ability and daring, these views now seem naive. To a great extent, the same fate was reserved for the proper scientific investigation of light that flourished particularly from the seventeenth century on, with the scientific revolution. Though the progress made was undeniable, it was marked by doubt and dispute. Light has proved highly resistant to comprehension: the moment one of its properties is discerned, another more important one appears; just when the end seems in sight, it draws further away again.

Light, then, is the basic key to comprehending the universe itself. This truth, which marks the entire course of natural science, reached its apotheosis in the twentieth century: the theory of relativity; the establishment of quantum physics; the discovery that the universe is expanding; laser beams and cosmic microwave radiation, to mention but a few of the important steps made by modern science, are rooted in the sustained effort to comprehend light.

At the same time, the technological culture that is both our pride and a threat to our existence has swathed every human activity in electromagnetic waves; and all over the planet it has created a communications and information web unimaginable only yesterday. Electromagnetic waves, too, are an invisible side to light. Equating the two was once again a wondrous step in a long, eventful quest.

Yet though light as a multi-faceted entity has come to dominate both everyday life and science, questions still remain as to its true nature. Familiarity with light and its colours leads to facile illusions. These are instructive nonetheless, as they reveal the simplistic ease often typical of the way we think. So there is no end in sight for the wonderful adventure of light, which began so long ago. It can not only boast of great moments and crucial errors, but also of the constant attempt to search further. It has led us to an understanding of the cosmos and revealed its hidden dimensions, but has also pointed out our limits, imposing a degree of humility on us. It is worth noting that in the closing years of his life, having triumphantly introduced photons as the fundamental "quanta" (grains) of light, Einstein himself admitted his inability to grasp their true nature.

In the broader sense, it is thus didactic to make at least brief reference to the history of light. This history cannot be isolated from the prevailing atmosphere in each era, or from its broader cultural coordinates. Indeed, quite the opposite. The history of light is bound up with the course of humanity and its achievements, fears, prejudices and anxieties. In his wonderful *Preface to the Treatise on Vacuum*, Blaise Pascal observes: "Thence it is that by an especial prerogative, not only does each man advance from day to day in the sciences, but all mankind together make continual progress in proportion as the world grows older, since the same thing happens in the succession of men as in the different ages of single individuals. So that the whole succession of men, during the course of many ages, should be considered as a single man who subsists forever and learns continually [...]"[1]

So light's gambols through history are gambols by humankind itself—and that is something that you, dear reader, should always bear in mind.

1.2   Flashes of Light in Antiquity

Long before it became an object of scientific enquiry, light and its sources were accorded divine status. The world's mythologies boast images and myths beyond compare, telling of the sun and the moon, the stars or fire. The sun was the source of life, symbolising the creative power of nature and acting as man's guide to the hours and seasons.

From very early on, light was a symbol that condensed all the elements of good in the world. To the Persians, for instance, Ahura was the god of light and represented beauty, wisdom and benevolence. Ariman, on the other hand, was the lord of darkness and the personification of evil. According to Egyptian mythology, the sun in the form of the god Ra was the creator of the world, who touched the Earth and gave birth to its elements. Of these, fire and moisture held pride of place. The stars were believed to be the souls of the dead kings, who ascended into heaven to become one with the Sun God or to join his retinue.

Later, in the more sophisticated Greek world, sun worship became associated with Apollo. Light as an entity had a significant presence in theories of cosmology. According to Hesiod's *Theogony*—the earliest depiction of the Cosmos, and one of the most poetic—the primordial trinity of creation comprised Chaos, Gaia (Earth) and Eros. Night and Erebus were born of Chaos, and from their union sprung their opposites, Day and Aether. This was an expression of the established pre-scientific knowledge that day succeeds night, and that darkness gives way to light.

The significance of light is also underlined by a beautiful image encountered in the work of the comic poet Antiphanes: "Of Night and Silence came Chaos, and then of Chaos and Night Love, and of Love Light, and of Light the first generation of the gods."[2] The idea that light derives from Love, and the gods in turn from it, would be difficult to dispute even today! In any case, the uniqueness of Greek light emerges from as early as Homeric times. "Only in Homer's Greece," notes Roberto Calasso, "does the cry of the warrior who begs Zeus that he may be killed in the light make any sense: 'Destroy us in the light, since such is your pleasure.' The light will not serve to escape death but to usher it in. A death in the gloom of the fog would already be a fragment of the sorrowful afterlife, all weakness and vacillation, whereas a death in the light is a last instant of clarity."[3]

Lastly, it is worth stressing that fire, which is both a tool and light, was what Prometheus chose as an invaluable gift to humans. And since it signifies the dawn of technical civilisation, the associated myth is timeless in its symbolism.

Nevertheless, any attempt to provide an account of ancient peoples' myths concerning the Sun and light seems futile and in vain, such is their wealth and imagination. And however much it boosts our ethnocentric sense of conceit, the truth remains that the first ideas concerning the true nature of light are encountered in Ancient Greece. They are philosophical in origin, but often contain elements of science. In any case, it is well known that Ancient Greek "philosophy" encompassed all branches of learning, and that philosophers were universal thinkers, free of compartmentalised knowledge. As the historian of science C. Gillispie stresses: "Of all the triumphs of the speculative genius of Greece, the most unexpected, the most truly novel, was precisely its rational conception of the cosmos as an orderly whole working by laws discoverable in thought. The Greek transition from myth to knowledge was the origin of science as of philosophy. Indeed, knowledge of nature formed part of philosophy until they parted company in the scientific revolution of the seventeenth century."[4]

The attempt to interpret the world via logical argumentation and reasoning had its beginnings on the shores of Ionia. Miletus, a city both wealthy and in contact with the cultures of the East, was in the sixth century AD a hub of innovative ideas that were rapidly disseminated throughout the Greek world. Great thinkers known as the Presocratics developed the first theories on the creation of the world and the nature of the senses, making frequent references to light and vision. As stressed by Constantine Vamvacas, a leading scholar of their works, "The goal of the Presocratics is ambitious. Their aim is to *comprehend* not to describe nature. They are so acute and visionary as to perceive that any involvement with partial, experimental details would necessarily mean diversion from their sole end, which is to *apperceive* the cosmos in its *wholeness*."[5]

It should be noted that very early on, Parmenides made systematic use of light and nighttime to account for physical phenomena and to build his own cosmological edifice. However, the few surviving fragments of his work are too insufficient and opaque to permit reconstruction of his complex cosmological system, which also included "annuli" or rings of light and darkness. The presence of light in Parmenides' theories is attested to by one of the most beautiful lines in Greek literature:

νυκτιφαές περί γαίαν αλώμενον
αλλότριον φως
"Astray over earth / Bright in darkness /
Its light also a wandering foreigner"[6]

which, in free translation, refers to the light that shines at night and wanders around the Earth, coming from foreign parts.

So the history of optics begins with works of natural philosophy. Empedocles, Leucippus and Democritus are pre-eminent among the Presocratics who concerned themselves with the nature of light and the operation of vision. They were all alive in the mid-fifth century, as the "Greek miracle" was approaching its apogee: the Parthenon had begun to rise into the light of Attica; the tragedies of Aeschylus and Sophocles were being performed in the theatres; and a humble man named Socrates was teaching in the Athenian Agora. As this brilliant era drew to a close, Plato was impatiently awaiting his moment.

The philosophical ideas of Empedocles were to have a major influence on his successors. He himself was a curious combination of philosopher, mystic and shaman. His ideas on the operation of vision overturned the previous Pythagorean view, which asserted that optical rays were emitted from the eyes and reflected off objects, creating their image. By contrast, Empedocles focused his attention on objects themselves: light rays "emanate" from their surface, conveying information about them. A kind of optical sensor extending from the eye then picks up the information and forms the object's image. In other words, there is a clear analogy with the finger, which feels a body and senses its shape. The theory of emanations was in any case trying to provide an interpretation for all the senses; and it argues that we feel something when it fits into the pores of each individual sense. It is also worth noting that according to Empedocles, sunlight is made up of particles, and arrives first between Earth and the heavens before reaching us. Its motion escapes us because it occurs at enormous speed.

The theory of emanations was accepted and elaborated by Leucippus and Democritus, the atomist philosophers. They held that under the influence of light, the surface of objects constantly generates extremely thin veils of matter, known as *eidola* or images. These move continually at immense speed in all directions. They do maintain their shape, however, and as they reach the eye at brief intervals, they create a sense of continuity in a changing image.

Given that the atomist philosophers endorsed the theory of emanations, it follows that they would have linked it to the idea of atoms, which was one of the most brilliant conceptions in Ancient Greek thought. So does a body appear red because it emits red atoms? The answer is negative. Colour does not number among the primary attributes of atoms, like their size or shape. The atoms of *eidola* simply touch our eye in such a way as to create the impression of red. "By convention sweet and by convention bitter, by convention hot, by convention cold, by convention colour; but in reality atoms and void,"[7] Democritus stresses. Thus, atoms themselves only have few attributes; and the image we see on receiving them is created by our own selves.

Perceptions of the role played by atomic particles in conveying the sense of hearing or sight remained dominant until the Renaissance. The fact that, a century after Democritus, atomic theory and the idea of *eidola* were accepted by Epicurus played an important role in their dissemination. His views are summed up in the following excerpt from one of his letters: "There are patterns of shapes similar to those of solid bodies but quite different from any appearance we know on account of their fine-grained quality. Because it is not beyond the power of nature to make such subtly refined lengths of space within the all-encompassing <empty space;> nor to make such skilfully fine-tuned arrangements that can craft the hollow and smooth surfaces; nor to make outflows maintain the same <coordinate> positions on two dimensions, which they had analogously when they were in the solid bodies. And we call such patterns images [*eidola*]. As their movement through space meets no countervailing resistance from bodies that could rebuff them, these images can complete any conceivable distance in an inconceivably short period of time."[8]

It would appear that the workings of a world that consisted of impersonal, neutral atoms was consistent with Epicurus' aversion to the abstract and the supernatural. "It is impossible to be released from fear about the most important things for one who, not having adequate knowledge as to what the nature of the whole is, is trying to second-guess this or that in accordance with the <traditional> fairy tales."[9]

Epicurean perceptions of both the senses and atomic theory are encountered at length in one of the famed epics of Roman times—Lucretius' *De rerum natura*, written circa 50 AD. The fact that it was written in Latin is of particular importance. The Greek world was in decline, and the Latin language began to gain wide currency in educated circles. According to Lucretius, light is a necessary condition for us to perceive *images*, which are formed of atoms. That light derives from the sun, the moon and various other "fires", and is a specific substance behaving in its own particular way.

The epic by Lucretius envisages the universe as a combination of atoms but also contains elements of evolutionary theory. Interestingly, it was greatly admired and constantly copied, possibly because Lucretius' materialistic and dialectic outlook and his Epicurean influences opposed Christian dogmatism, which had begun to gain ascendancy.

That being said, it is worth noting that just like human history, the history of science has often been hijacked by dogmatic views. Fortunately, once experiments became established, science found a way to outflank them. Testing ideas by experiment is an incontestable strength of science, and a hallmark of its progress. Dogmas and revealed truths have always been a source of human suffering and arrogance.

1.3   The Platonic View of the World

In the second half of the fifth century BC there was a marked shift in philosophy, towards the human individual and the problems of life. This radical shift articulated the reaction of common sense thinking against the world as presented by natural philosophers, which was remote and unintelligible. The ascendancy of a new class—the sophists—who earned their living by guiding people in practical matters, was the result of this more general human-centred shift. What set the sophists apart was a scepticism bordering on irony with regard to absolute values and knowledge.

Into this transitional world stepped Socrates, who attempted to restore the damaged credibility of philosophy and moral values. On the other hand, his gifted pupil Plato made it his life's work to disseminate and advance Socratic teaching and its moral lessons, which had hitherto been oral. Following the death of Socrates and extensive, adventurous peregrinations, Plato returned to Athens. There he founded a kind of school of philosophy (Fig. 1.1), the famed Academy. He wrote around thirty dialogues centred on metaphysics, ethics and politics.

It is characteristic that, as the leading twentieth century philosopher Alfred Whitehead believed, all of Western philosophy after Plato could be read as mere footnotes. For all its apparent hyperbole, this comment reflects both the historical weight and enduring quality of Plato's work. Rather than acting as an authority, he motivates people to search for the truth, asking successive questions and attempting to upset pre-existing certainties. He thus chooses Socrates as the central figure in his dialogues, not simply by way of reference to his teacher, but so as to reveal the delusion of the senses and phenomena via the Socratic method known as maieutics. In unparalleled fashion, his work reveals the dialectical course to eternal, archetypal ideas of goodness, beauty and truth, which are only encountered as imperfect imitations in the everyday world.

Yet as far as light and the operation of vision are concerned, Plato has little to add to that held by Empedocles and the atomist philosophers. His account represents a fuller synthesis nonetheless, and he believes that the light issuing from the eye is of equal importance to the light of the sun.

Fig. 1.1 Rapahael's famed fresco *The School of Athens* (1511) in the Apostolic Palace, Vatican City, representing Greek philosophy (Vatican City Museums/PD)

According to Plato, a gentle light due to the internal fire in the eye coalesces with the light of day, thus forming a homogeneous luminous substance. This substance constitutes the bridge permitting images from the outside world to reach the soul. Consequently, there is a deeper harmony between the sun and functioning of the eye. Centuries later, the same idea is reflected in Goethe's lines:

Were not the eye itself a sun,
No sun for it could ever shine:
By nothing godlike could the heart be won,
Were not the heart itself divine.[10]

Everything related to the senses is encountered in the *Timaeus*, one of Plato's later dialogues, where he uses poetic language to deal with the creation of the world and describe its fundamental structures. As he stresses in characteristic fashion: "And of the organs they [the gods] first contrived the eyes to give light, and the principle according to which they were inserted was as follows: So much of fire as would not burn, but gave a gentle light, they formed into a substance akin to the light of every-day life; and the pure fire which is within us and related thereto they made to flow through the eyes in a stream smooth and dense, compressing the whole eye, and especially the centre part, so that it kept out everything of a coarser nature, and allowed to pass only this pure element. When the light of day surrounds the stream of vision, then like falls upon like, and they coalesce, and one body is formed by natural affinity in the line of vision, wherever the light that falls from within meets with an external object. And the whole stream of vision, being similarly affected in virtue of similarity, diffuses the motions of what it touches or what touches it over the whole body, until they reach the soul, causing that perception which we call sight."[11]

As is evident from other fragments of Plato's oeuvre, as a philosopher he believes the operation of the senses to be supported by an emanation mechanism. The sense organs have microscopic pores and, since sound or colour particles are of different shape and size, they penetrate depending on whether the organ is designed for sight or hearing. Consequently, the mechanism for sight is complex; it needs to link those particles to the fire emanating from the eye, decode messages about colour and shape, and convey the information to the soul. According to Plato, it is thus in a position to create illusions or delusions. The rainbow is a similar illusion—it does not exist in reality, just as the bizarre images in a fairground mirror do not exist. Only the eternal, immutable world of forms, which has the soul as its eye and God as its light source, is the incontestable foundation of truth.

It should be noted that Plato's god is artistic in nature, sculpting the physical world on the basis of an ideal model whose perfection he strives to approach. What is more, the status the *Timaeus* reserves for human beings is impressive. In a sense, they are the universe in miniature. It follows that if one synchronizes one's stance with the harmony governing the universe, then one's actions will be rational and moral. Not being restricted to the mundane needs of everyday life, the task of vision within that framework is characteristic: "This much let me say however: God invented and gave us sight to the end that we might behold the courses of intelligence in the heaven, and apply them to the courses of our own intelligence which are akin to them, the unperturbed to the perturbed; and that we, learning them and partaking of the natural truth of reason, might imitate the absolutely unerring courses of God and regulate our own vagaries."[12]

Plato's theory of the senses is no more than a minute part of his philosophical oeuvre. The relatively extensive reference made to it has another reason beyond the special significance he bears as a philosopher. Up until the eleventh century AD, the *Timaeus* was the only Platonic dialogue translated into Latin, and it had an enormous impact on the Christian and subsequently the Muslim world. Among other things, it thus contributed to the dissemination of Presocratic ideas. The creation of the world by a beneficent divinity was in any case in line with Christian dogmas.

The powerful influence exerted by Plato's oeuvre on Christian thought is thus no coincidence. Typical of this is the fact that eight centuries later, in his treatise *On Genesis*, Saint Augustine was to repeat what Plato had maintained on the operation of vision. As he concludes: "But still, the light which is in the sense of the seeing subject is so slight, we are informed, that unless it was assisted by the light outside, we would be able to see nothing…"[13]

It should be noted that everything known in the fourth century BC about medicine and astronomy, psychology and the natural sciences is lucidly summarised in Plato's *Timaeus*. Today, at a time when scientific knowledge largely appears sterile and specialised, the Platonic dialogue can still teach the need for a unified view and a comprehensive outlook on the world.

1.4   The Brilliant Ascendancy of Aristotle

In Ancient Greece, the archetypal, transcendental significance of a pupil teacher relationship often came to the fore. Thus Plato recorded and commented on Socratic teaching, but also broadened it. In turn, as a pupil in Plato's Academy, Aristotle not only criticised his teacher's ideas; to a great extent he overturned them, so as to found his own edifice. In that way he was to become the second leading thinker in antiquity to rapidly transcend the boundaries of his own time and place.

The interesting thing is that the bond between teacher and pupil was to have its sequel, but this time in a paradoxical manner. Born in Stagira, Macedonia in 384 BC, Aristotle spent some time as a private tutor to Alexander, son of Philip of Macedon. When Alexander gained the throne, Aristotle made good use of his pupil's support in founding his famed Peripatetic School in Athens. There he was to teach until Alexander's death, leaving a monumental oeuvre in the form of students' notes. His work touched on almost every area of scientific enquiry: medicine, physiology, meteorology, physics and psychology, as well as issues in biology or natural history. Aristotle established the systematic study of logic, while his *Ethics* is a work unsurpassed. He also made a significant contribution to political theory and literary criticism.

As far as the operation of vision was concerned, Aristotle remained unconvinced by previous interpretations, which either held that light was particulate radiation or that the eye emitted optical rays. More generally, he was not keen on atomic theory, as it was irreconcilable with his ideas on the primary qualities inherent in things. Nor was he convinced by the way Empedocles and Plato accounted for our inability to see at night. As Aristotle rejoined, "The explanation in the *Timaeus*, that the sight issuing from the eye is extinguished in the darkness, is quite without point, for what can the extinction of light mean?"[14] He himself accepted the need for a physical medium between the object and the observer, a medium that must be transparent. It is thus not something we see, but something via which we can see. It is filled with images, which are conveyed to the senses in some way.

It follows that Aristotle regards light as a state of the transparent medium, brought about by the presence of fire or another luminous body. This explains why it does not take time to disperse, since it is a "state" rather than a substance. Lastly, as the philosopher explains in his work *De sensu et sensili*, colour is a characteristic of visible objects that is capable of setting the transparent medium in motion. Black and white are the fundamental properties of colour; mixing them on a surface produces all the other colours.

In essence, the Aristotelian theory of vision provides us with a reference vocabulary for the operation of the senses. Its virtue lies in the fact that it forms part of an extremely broad system that attempted to interpret the natural and spiritual world in a consistent manner. It is thus no paradox that the theory's chief characteristics are encountered in Aristotle's work *On the Soul*, a majestic treatise on the sum total of human nature.

Turning to the four fundamental elements proposed by Empedocles—water, fire, earth and air—it is interesting that Aristotle felt the need to add a further one, the quintessence. Stable and immutable, the quintessence somehow links the other four elements, which are in constant flux and which compose things. Indeed, inspired by the constellations, which remained immutable throughout their eternal courses in the heavens, Aristotle held that the quintessence was made of the matter of the stars. This hypothesis "was close to the reality of twentieth-century physics," as surgeon and author Leonard Shlain observes. In his book on the relationship between Art and Physics, he goes on to say: "The Quintessence [...] is not the stars, but rather light itself. This too is fitting. Fleeting and enigmatic, this fifth essence has engendered wonder and reverence throughout history. Whether it was the miracle of fire or the life-giving rays from the sun, light in and of itself has always been the most mysterious element. It has been accorded a prominent place in all religions, and discoveries in modern physics revealed that it was the unique nature of light that held the key to unlocking the secrets of the other four."[15]

It is also worth noting that Aristotle was the first to make a serious attempt to understand rainbows and how they are created in the rain-filled sky. To that end, and despite the reservations he had expressed, in his *Meteorology* he postulates the existence of a kind of optical ray issuing from the eye. He then goes on to observe that only colours—but not the images of objects—are reflected in small bodies, as in dewdrops sparkling in the sun, for instance. He thus surmises that the optical ray leaves the eye and bounces off the raindrops in clouds, before moving in the direction of the sun, behind the observer. His interpretation is enriched with many details, bearing the first recognisable traces of a scientific method.

Over the centuries that followed, Aristotelian thought was to have an enormous impact. After the philosopher himself died in 322 BC, dissemination of his word was assisted by the Neoplatonist School, which flourished during the dying days of the Roman Empire. In the Middle Ages, Aristotelian ideas are encountered in the tenets of Islamic philosophy; they were mainly incorporated into Western theology by Thomas Aquinas. Aristotle's influence is still being felt to this day, and his ideas are always heavily present in the world of thought and academic debate.

As our references to the history of light at the time of the Greek miracle draw to a close, it is vital that one self-evident truth be underlined: everything mentioned here, of necessity only in brief, should only be regarded as mere traces of a world rich in ideas and intellectual accomplishments. Yet as those traces are presented piecemeal and in isolation, they often lose the tie binding them to the creator's entire oeuvre. Only an entire reading of Plato's *Timaeus* or Aristotle's *On the Soul* could restore the proper weight of a reference to the operation of vision.

Be that as it may, it is interesting that when doing science, the Greeks avoided quantitative calculations or at least some simple experimental proofs. "Why the Greeks, for all their brilliance of intellect, made at this time so little use of experimental methods," writes W. K. C. Guthrie, "and no progress at all in the invention of apparatus for controlled experiment, is a complicated question. Aristocratic tradition and the presence of slaves no doubt had something to do with it, but are scarcely in themselves a sufficient explanation. To some small extent those in the Ionian succession made use of observation, but only spasmodically until the time of Aristotle, and of controlled experiment they had no idea. Their legacy lies elsewhere, in their astonishing powers of deductive reasoning."[16]

It is also worth noting that the interpretation of vision first introduced by Empedocles, which presupposed the existence of some kind of optical rays, lived on in various different versions for centuries. A parallel course was charted by the complementary and partially rival theory supported by the atomists, which attributed vision to images that fly in space until apprehended by the soul and the eye. Despite the authority of its originator and the fact that it did not in essence conflict with the existence of optical rays, the Aristotelian notion of a transparent medium had little impact on things. It is too well rooted in the realm of the abstract, integrated into the great philosopher's overall attempt to arrive at a satisfactory understanding of the senses.

From one perspective, this was a true stroke of luck for the history of light. It well known that in science as well as in life, luminaries carry the rest of us along with them, not only in their great ideas, but also in their errors.

1.5   Geometric Light

While none of the earlier philosophers directly pose the question as to what light is, it surely does not cease to preoccupy them; and implicitly speaking they do provide some answers in their theories of vision. The closest to some kind of answer—though it too remained up in the air—was Aristotle, when he regarded light as an "action" (*drasis*), a kind of state of the transparent medium. The crucial question was subsequently left aside once more, but there was a substantial development: investigations began into the behaviour of light rays, their reflection off smooth surfaces, the way they moved and the power transmitted by them. In any case, the sun sent humans warmth and light, the moon illuminated the night and also caused the tides, and the stars appeared to shine eternally and rule the fate of human beings. So light was not simply linked to vision—on the contrary, it was one of the world's main constituents.

The first person to put forward some form of mathematical data on the behaviour of light was Euclid, the great geometer of antiquity. Though much is known of his work, the opposite is true of his life. He was probably a student at Plato's Academy, and in around 300 AD moved to Alexandria, the new city built by Alexander the Great in Egypt, which grew into a brilliant centre of Greek culture. Euclid systematically assembled all the mathematical knowledge of his time, lending it a wonderful unity founded on definitions, axioms and theorems. Nine of the thirteen volumes in his *Elements* are concerned with geometry, and the remainder with number theory. In the *Optics* his attention is particularly focused on vision, and his views are essentially Platonic: a bundle of optical rays issues from the eye, and in some way senses the objects it encounters. Euclid maintains that these rays are discrete, and are of the same nature as light rays. However, their behaviour obeys specific geometrical theorems and axioms. Euclid formulated the fundamental principle of geometrical optics, i.e. that light travels in a straight line, and is reflected at the same angle it hits a surface. He also describes the change an optical ray undergoes in refracting materials, though without arriving at arithmetical calculations. The significance of his *Optics* lies in the fact that it treats phenomena actually observed in nature—such as light and the operation of vision—with the stringency of mathematical logic.

Euclid's path was trodden by yet another thinker, who lived in Alexandria almost two centuries later, and who was to influence the evolution of science as few others: Claudius Ptolemy. Founded on the cyclical motions of the planets and the sun, his admirable geocentric system (Fig. 1.2) permitted accurate predictions to be made about their positions and was to survive up until the seventeenth century and the Copernican revolution. As the eminent scholar of antiquity Vasilis Kalfas notes, "Ptolemy's astronomy bears no substantial relation to the convictions of his contemporaries or predecessors. His method rests on acknowledging that mathematics is the key to approaching celestial phenomena, while its sole functional criterion is the principle of simplicity."[17]

Ptolemy also did significant work in other scientific disciplines, such as geography and optics. The *Tetrabiblos* is a lengthy treatise on astrology and planetary influence on human affairs, and was long held to be the most authoritative source on the subject. It is not devoid of scientific elements,

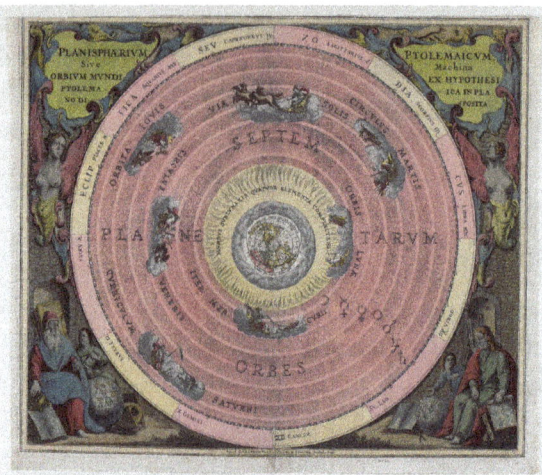

Fig. 1.2 The geocentric "Planisphere of Ptolemy", plate 1 from Harmonia Macrocosmica (1660) by Dutch-German cosmographer Andreas Cellarius (Minneapolis Institute of Art/CC PD)

however, and in some way is a supplement to the *Almagest*, which contains Ptolemy's theory and calculations on the movements of the planets. The *Optics* only survived in part, and that in Arabic translation.

It is nonetheless significant that Ptolemy not only extended Euclid's mathematical analysis of optics, but also incorporated elements of physics, physiology and psychology into it. He attributed vision to the action of optical radiation extending in a cone from the observer's eye. This radiation, he believed, was of the same nature as external light, and transferred energy, which decreased on reflection or on entering transparent media.

Yet Ptolemy's most valuable contribution with regard to light lies in the experimental methods he employed. The great Alexandrian thinker really did record the actual angles of incidence and refraction of a light beam on passing through a transparent medium. He thus arrived at satisfactory approach to the law of refraction. What is more, he had ascertained that in order to determine the precise position of a planet he would have to bear in mind the curvature of the light beam on entering the earth's atmosphere—so a star or a planet appears to be higher in the sky than it actually is. The refraction of sunlight is what appears to raise the sun and extend the duration of the day.

It should be stressed that while the theoretical study of light preoccupied the thinkers of the Greek world and elsewhere, there were also extremely wide-ranging applications based on the associated conclusions and relevant experience. Precise knowledge of the sun's movements already existed in

Ancient Egypt, a fact brought to the fore by some of its wondrous monuments. Besides, as Homer recounts in the *Iliad*, the goddess Pallas Athene helped Diomedes to dazzle his adversaries with sunlight reflected from his shield and helmet. Using mirrors in a similar way, Archimedes focused the sun's rays and set fire to enemy ships. And in contrast to buildings in Greece today, the masterpieces of ancient architecture took into consideration the routes taken by light and its games with shade. Care over proper orientation is even evident in ancient theatres. Thus, the light conditions prevailing during the Great Dionysia festival in Athens allowed people to attend theatre performances from morning to afternoon.

At around the same time that Euclid and Ptolemy lent the investigation of light concrete, would-be experimental content, a religious movement named Manichaeism reduced light to the realms of the absolute and a symbol of Good. This movement appeared in the early second century AD, founded by the Persian prophet Mani. The interesting thing was that it lasted for over a millennium, flourishing in Rome, Africa and even China. Manichaeism brought together elements of Christianity, Zoroastrianism and Buddhism, but its basic constituent was the absolute dualism of Light and Darkness. It maintained that the origin of light was good, whereas the origin of darkness was evil. Because the rays of Light had become trapped in the material world, a holy war broke out between the powers of Light and those of Darkness. In the course of this war, the Prince of Light created the First Man, who was, alas, taken prisoner by the forces of Darkness and remained captive to matter. The creation of Adam and Eve was also the work of the Prince of Darkness, but at the same time he took pains to imprison the elements of light in their material bodies. Consequently, the only process of redemption open to contemporary humans was the knowledge that their soul contained elements of the good God's nature. This was the truth revealed to Mani, which formed the core of his prophetic teaching.

It is worth stressing that the followers of Manichaeism had no temples, images or other visible elements of worship. Yet neither that fact nor the indisputably good intentions of Mani's religion prevented the Persian authorities from sentencing him to death in around 276 AD. As usually occurs in such instances, his conviction had precisely the opposite of the desired effect: the Song of Light continued to be heard long after, and despite provoking reactions from every form of orthodox belief, had a broader impact on religious understanding.

Interestingly, the Song of Light can still be heard to this day, loud and clear and rich in harmonies; but its motivator and composer in present times is science, which has revealed that light is the principal force in the workings and comprehension of the world.

1.6   Light in the Thousand and One Nights

From the time of Euclid and Ptolemy, more than eight centuries were to elapse before the history of light saw any noteworthy peaks again. In the meanwhile, the world had changed radically. The faithful soldiers of Islam had built an entire empire extending in the eighth century from the Mediterranean to the Indian subcontinent. Islam was to prove tolerant to the cultivation of knowledge, and indeed often encouraged it. There was thus a flowering of science and philosophy, founded on Ancient Greek thought and its major intellectual achievements.

One highly gifted and cultivated personality, Harun al-Rashid, the Caliph of Baghdad, was to make his mark on contemporary affairs and support for the arts. The ancient authors were zealously translated, and a conscientious effort made to disseminate them. The city of Baghdad was thus to experience a brief golden age. Poetry and music flourished, and echoes of the prosperity and atmosphere of the times still ring through the marvellous *Thousand and One Nights*. The cultivation of learning was not to die out completely even in the wake of Harun's death, the rivalries between his successors and destructive raids aimed at capturing the glittering city's wealth. However sad it is to say, the city was to suffer one of the most barbarous raids in recent times, too— and that in the name of the ideals of democracy and freedom.

Be that as it may, the blossoming of intellect that accompanied the ascendancy of Islam, particularly in the centres of the Middle East and the Mediterranean, was to yield highly influential thinkers. Among them was the Persian mathematician al-Khwārizmī, who lived in Baghdad and worked mainly on algebraic calculus; the doctor and philosopher Avicenna, who had a significant impact on later thinkers; Averroes the philosopher, a profound scholar of Aristotle and Plato, who insisted that philosophical truth must be founded on discourse, thus reconciling Greek and Islamic thought; and lastly, Alhazen, an outstanding physicist, who was to strive towards a major breakthrough in his theories of light and vision.

The first philosopher in the Islamic world to work on optics was al-Kindi. He lived in Baghdad in the ninth century, and wrote books on science, philosophy and music. Although a work of astrology, his *De radiis stellarum* (*On Radiations from the Stars*) digresses into philosophy, devoting little space to predictions or practical advice. According to his theory of vision—which is basically Ptolemaic—optical rays are shaped like cones and link our senses to the world. Under their influence, air is rendered capable of transmitting the shape and colour of objects. Interestingly, however, al-Kindi believes that radiation travels in all directions, and not only issues from the eye. "It is manifest," he writes, "that everything in this world, whether it be substance or accident, produces rays in its own manner like a star".[18] A web of rays thus covers all of space, linking the Earth, the stars, fire and magnets. It follows that our eyes receive rays wherever they are. Even sound is a form of radiation, but there are a host of other forms not picked up by the senses. These radiations, which al-Kindi calls "metaphysical", cause mysterious effects. Since each of the stars is of differing composition and at a different distance, they act on objects in various ways.

The universe surely seems incomprehensibly complex when suffused in radiations of every kind, each with a different task and effect. Yet that does not appear to concern al-Kindi. As a philosopher, he believes there is a deeper harmony that is the work of God, to which humans have simply to adapt their free will.

Roughly a century later, a village south of Baghdad was the birthplace of Abu Ali al-Hasan ibn al Haitham, who was to become known in the West by the Latinized name Alhazen. Very little is known of his life. At some point it appears that he settled in a Cairo mosque which was a kind of theological university, and that having promised but failed to control the flooding of the Nile, he feigned insanity to escape the wrath of the caliph. Alhazen not only translated Euclid and Ptolemy, but also wrote around 120 books of his own. His work on optics is monumental, and overturned prevailing views. Despite being translated into Latin in around 1200 AD, it was not widely known in Europe for a considerable time, possibly on account of its bulk; but on gaining currency it was to dominate teaching on optics up until the seventeenth century, when the scientific revolution blazed new trails.

The difference between Alhazen and the Greek philosophers and physicists who worked on light is immediately apparent. In place of axioms and hypotheses, Alhazen persistently took an experimental approach, while at the same time relying on algebraic and geometrical applications. With the aid of flat and curved mirrors and lenses, he was to study the reflection and refraction of light in detail. He also formulated a hypothesis only confirmed in the nineteenth century—that light has greater difficulty travelling through denser materials, and that refraction is due to this.

That aside, one of Alhazen's crucial findings was that light rays travel in straight lines. He confirmed the truth of this by observing the light coming from the sun, the stars and distant flames. Furthermore, he believed that colour and light were inseparable, being radiated in all directions from every point or small area on a body's surface. And since rays are linear, they reach the eye as they set out, without interfering with each other. As for the nature of light, Alhazen contended that light was an occasional attribute of an illuminated object, which in turn became a source of light itself. Indeed, the dissemination of light was secured that way—every point in the air was illuminated to become a new source of light travelling in every direction. This radical conception foreshadowed Huygens' wave theory, which was to gain ascendancy many centuries later.

Alhazen's thinking obviated the need for the optical ray that had hitherto dominated theories of sight. "There is no vision," he writes in his *Optics*, "unless something comes from the visible object to the eye, whether or not anything goes out."[19] He thus relied on experimental observations and a handful of hypotheses to work out a comprehensive, astoundingly detailed theory of vision, not far removed from current perceptions.

One fundamental constituent of Alhazen's theory is that via the light it emits, every point on the surface of an object is depicted on the eye's crystalline lens. A true but small-scale image is thus formed, and then transmitted to the soul. This process is clearly far from simple: "Many visible properties are perceived by judgement and inference in addition to the sensing of the object's visible form."[20] It should be noted that Alhazen built a camera obscura and experimented with candles and their images on a screen in order the confirm his views. But it is only fair to point out that the optical principles of the camera obscura had already been mentioned in the work of Aristotle, and are also encountered in Chinese texts from the fifth century BC.

It is obvious that both Ptolemy and (even more so) Alhazen were the first to study light in what could be termed a scientific manner. Their predecessors were either philosophers at pains to create a comprehensive construct of the senses and the world, or mathematicians such as Euclid, who used geometry to describe the path taken by optical rays. Alhazen did not reject the geometrical method, but he reversed the trajectory of the ray, making it head from the object to the eye. The hitherto fuzzy role played by light suddenly acquired a sharper content. Light is a substance. It comes from the Sun, or from a flame, and touches objects. So, for instance, a vase situated opposite us receives the luminous substance, which spreads from each point on its surface in all directions, while simultaneously conveying the form of its colour. Thus anyone who observes the vase will receive the light and the colour entering their eye to form its image. From there, the image will be transmitted to the soul, which remains the ultimate authority and judge.

Readers may be puzzled by the significance lent here to Alhazen and his theory. Nevertheless, the fact remains that the theory represents a true watershed in the understanding of light; indeed, it foreshadows the scientific revolution several centuries ahead of time. But there is another reason for the extensive reference made here. It is common knowledge just how much Western education abounds in admiration and detailed description of every European intellectual achievement, and how much our universities offer in-depth study of the people who have contributed to scientific progress. Yet little attention is paid to the ideas and cultures that developed elsewhere, even if they were prime contributors to present-day knowledge. Figures such as Avicenna and Alhazen far outlived their time, and present-day humanity would look very different if Arab thinkers had not studied ancient texts with such affection and perseverance, and had not gone to such lengths to disseminate them.

1.7   Bursts of Light in the Middle Ages

It is common knowledge that the ancient world which grew up around the Mediterranean Sea became politically unified under Roman rule. Culturally, however, it remained strongly rooted in Greek thought and art. After the breakup of the Roman Empire in the fifth century AD, medieval Europe was to emerge in a hazy, uncertain manner, extending beyond the bounds of Roman territory. Up until the fall of Constantinople in the fifteenth century, the Byzantine Empire was a powerful presence in this new landscape, and its distinguishing features came to the fore in customs, political institutions and culture. But what hallmarked the times was the growth of Christianity, with a worldview at heart foreign and inimical to classical antiquity. By and large, medieval scholars had institutional ties to the church and were often unsuccessful in their attempts to combine Graeco-Roman tradition with Christian teaching. As Edward Grant observes, "In its most meaningful sense, the history of medieval science is the history of the dissemination, assimilation, and reaction to Ancient Greek science as it passed from the Byzantine Empire to Islam and subsequently to Western Europe."[21]

To some extent, the history of light followed this movement of ideas. In schematic terms, the light ray began with Empedocles in Sicily and, thanks to Euclid and Ptolemy, revealed its mathematical structure in Hellenistic Alexandria. It later moved to Baghdad, where it assumed an Arabic identity. Indeed, Alhazen reversed its trajectory and enriched it with experimental observations. On returning to Europe, the light ray gained a Latin identity, and so was to continue on its way up until the scientific revolution. A different light—this time internal and cosmological—was also reaching medieval Europe by different routes, having set out from Plato and Aristotle.

Up until the mid-thirteenth century, when he departed this life at a ripe old age, one characteristic presence was that of the English scholar and bishop Robert Grosseteste. His prolific oeuvre included both scientific treatises and translations of Aristotle. Grosseteste may be regarded as the forerunner of the empirical school, since he argued that Ancient Greek theories should be tested by practical experiment. As a bishop he commanded great respect for his moral rectitude, and he attempted to implement a broad reform programme in the church. It comes as no surprise that he came into conflict with the Pope when reacting against corruption. In a letter shortly before his death, he was to conclude: "Therefore, reverend Sirs, on account of the duty of reverence and fidelity I do not obey; I resist; I rebel."[22]

Of all Grosseteste's books, pride of place is taken by *De Luce*, or *On Light*. An important work of Christian theology, it is heavily marked by the presence of Platonic philosophy—especially of the *Timaeus*—though not devoid of elements typifying modern science and thought. The basic idea in this treatise is that the entire universe originated in the diffusion and transformations of light. As Margaret Wertheim writes: "Thus, Grosseteste believed light was the key to understanding the working of the physical world. [...] Grosseteste concluded that a mathematical understanding of light would serve as the model for understanding all natural influence, or what we would now call force. In contemporary physicists' quest to understand the forces of nature, it is light that has generally served as the model."[23]

So according to Grosseteste's world view, light originated from a single point and began multiplying in all directions. It thus formed an ever-expanding sphere, while matter appeared together with it. The condensation and rarefaction of light then led heaven to separate from the earth, whence derived the seven heavenly and four mundane spheres.

As Grosseteste sees it, light is the primordial corporeal form from which the material world derives. Nevertheless, the sensual world is only one pole of creation. There is also another, supersensual world consisting of angelic light. It follows that the divine command: "Let there be light!" is of dual content. One aspect of it concerns our physical existence, which is due to the condensation of light into forms of matter. The second aspect concerns a metaphysical light of intelligence contained in God's spiritual creations. God Himself has Platonic attributes, having created the world with geometrical shapes and numbers as His yardstick.

Grosseteste was also the first person to acknowledge the fundamental role of mathematics in emergent science. "Nothing magnificent [in the sciences] can be known," he stresses, "without mathematics."[24] And although the archetypical, primary significance of light radiation had of course already been brought to the fore by the Arab philosopher al-Kindi, the English bishop was the one to incorporate it into a wondrous cosmological synthesis. Summing things up, Arthur Zajonc writes: "Robert Grosseteste is for us a figure who stands, Janus-headed, facing in two directions. One face gazes upward and back, evoking a metaphysics of light that embraces hosts of angelic beings and emanations of light that are active in God's creation as intermediaries and enactors of His will. His other face gazes earthward to a future when the physics of light will develop to its fullest extent as modern optics, and most especially as mathematical physics grounded in experimental observation."[25]

It was through the Church, though at the same time contrary to its dogmatism, that yet another figure of the times was to emerge: Roger Bacon. He is of particular interest not only for forming a distinct, structured theory of light, but because he was a true revolutionary in both thought and life. His creed appears to have been: "Let us stop submission to dogma and authority, and come to know the world!"

Bacon was born in England in 1214, and studied in Oxford and Paris, where the study and teaching of Aristotle were flourishing. He later joined the Franciscan Order and returned to Oxford, steadily gaining a reputation for his unconventional convictions and the breadth of his interests, which apparently ranged from the biological and physical sciences to alchemy and magic. He soon came into conflict with the leadership of the Order, who forbade him from publishing his views. All the same, his main ambition remained to compose an Encyclopaedia of Knowledge, so upon the intercession of a friend he sent a large portion of his work in secret to Pope Clement IV. For all his expectations, however, the response from the supreme ecclesiastical authority was a characteristically deafening silence. Indeed, very soon thereafter Bacon was sentenced to several years in prison for supporting newfangled ideas. Shortly after his release in 1292 he was to escape Church persecution once and for all, by departing this world.

Despite censorship and supervision, Bacon published numerous treatises on mathematics, logic and philosophy, the value of which was recognised much later. What is more, he had no hesitation in making prophetic reference to flying machines, the construction of telescopes, microscopes and the possibility of circumnavigating the globe, as well as to mechanical means of transport on sea and land.

Bacon attached particular importance to the study of light. This was not unusual among Christian intellectuals of the time, even if, like him, their ultimate aim was the purity of theological discourse. The reason for this was that we come to know the world, which they held to be the work of God, via the intercession of light and vision, hence optics is a fundamental science.

Yet beyond this elementary faith, originality is not the hallmark of Bacon's work. He follows the trail blazed by Alhazen and Grosseteste, aiming in some cases at synthesis, and in others at fine tuning previous ideas. Nonetheless, the interesting thing is that his optics are accompanied by mathematical diagrams and experimental description, thus rendering them a classic reference work over subsequent centuries. His major contribution lies in his general insistence on the value of experimental method, and the need for mathematical proofs. In that sense, Bacon was well ahead of his time.

Concisely termed the Middle Ages, that era was in any case slowly but surely giving way to the Renaissance and the scientific revolution. People were being liberated from theocratic beliefs, and gaining an awareness of their individuality. In the new political and social conditions taking shape, the role of science was to prove dominant.

1.8   An Astronomer on Supplements

We have seen how Plato and Aristotle's influence remained strong in the Middle Ages. To some extent, the philosophical works of that time were commentaries or extensions of their oeuvre. Within those terms, vision was treated more as an activity of the soul or spirit than as a physical process. But by the dawn of the sixteenth century the time looked ripe for a substantial shift, and the way to the scientific revolution lay open. This shift was neither homogeneous nor universal; but what had begun as a spiritual experience, whether relating to vision or light itself, slowly but surely acquired a fixed content, gradually separating out into optics and physiology.

Of course, the roots of this transition are identifiable in the work of Euclid and his ingenious geometrical conception of vision. But the interesting thing is that the new notions of light in the sixteenth century echo a radical shift in historical awareness, in a foretaste of what was to be a scientific outlook on the world. This shift was expressed par excellence by Kepler, Galileo and Descartes. All three were almost contemporaries, though the major differences in temperament that set them apart were to hallmark their reactions towards the all-powerful Church.

With Kepler, astronomy entered the emergent world of science once and for all. Copernicus had gone before—a Polish astronomer and cleric, who had dethroned the Earth from the centre of our planetary system a century earlier, putting the Sun in its place. It is, however, plain as the sun at midday that the idea of a heliocentric planetary system—daring in its time—was first conceived by Aristarchus of Samos. But he was harshly criticised by his contemporaries, and did not insist on his heretical views.

Be that as it may, it was Kepler who confirmed Copernicus' heliocentric views, and formulated the laws named after him on the motion of the planets. He believed their orbits to be mathematical ellipses, with the Sun at one of their foci. Kepler accurately calculated their orbital period and other elements of their motion. Once Newton appeared, however, the laws of planetary motion would turn out to be a striking consequence of universal gravitation.

Though Kepler is regarded as the founder of modern astronomy, it is worth noting that he was a deeply religious man. All the same, he held Ancient Greek thought in high regard, and Platonic ideas were to greatly influence his scientific make-up. Born in a small town in Germany in 1571, he first studied theology, but from early on showed a considerable flair for mathematics; his impressive career began as a mathematics teacher in Graz. His duties there included publishing an almanac that drew on astrological data to make predictions about the weather and days favourable to certain human activities. Beyond the fact that Kepler earned a comfortable living in this barely scientific manner, it should be pointed out that in many ways astrology still went hand in hand with astronomy at that time. In any case, Kepler regarded practical astrology and the use of the zodiac as fraud—his own astrological predictions were based on natural principles, as well as on his conviction that the natural world was unified. Indeed, in contrast to Galileo—who had assured a grandee of the time that he would live to a ripe old age, only for fate to strike a few days later—Kepler was lucky: his first astrological guide predicted storms and a raid by the Turks. Strangely enough, both occurred, and he gained a great reputation. Furthermore, his first studies on the planetary system led to positive comments from Tycho Brahe, the pre-eminent astronomer of the time, and it was on his encouragement that Kepler moved to Prague. When his benefactor died, he succeeded him as official Imperial Mathematician.

Kepler's work in astronomy is exceptionally wide-ranging, set down in dozens of books and treatises. His goal was the unification of mathematics, physics and astronomy. Light and optics are mainly dealt with in his book *Paralipomena* (*Supplements*, which appears in Latin in the convoluted title), where he comments on previous theories and advances on his own ones.

His main contribution, however, lies in understanding the mechanism of sight. Hypothesizing that the eye is a kind of *camera obscura*, he offers a detailed description of the course taken by light rays and how images are formed on the retina. His attention is drawn to the fact that according to this course, images should appear inverted, which does not of course occur; but he correctly regards interpretation of the phenomenon as belonging to physiology. His preoccupation with lenses was similarly dogged. Though mention is made of them as early as Aristophanes, their use was for the most part limited to improving eyesight. Galileo was to bring the importance of lenses to the fore by constructing one of the first telescopes and making his sensational discoveries in the sky. But it was Kepler's work *Dioptrice* that offered a theoretical interpretation of how images are formed by lenses and the way a telescope works. In an impressive manner, the scientific era of astronomy had already dawned.

Although the nature of light itself concerned Kepler less, Aristotle's influence was to lead him to an erroneous appraisal of planetary motion. Every movement, Aristotle contended, resulted from the exertion of a force. In turn, Kepler was to seek the particular force that moved the planets in sunlight, as he stood convinced that no angelic intelligence could bear that burden. "Although the light of the Sun cannot be the moving power itself," he writes in his *New Astronomy*, "I leave it to others to see whether light may perhaps be a kind of instrument or vehicle, of which the moving power makes use."[26]

Had Kepler been alive in 1638, he might have puzzled over the problem longer; it was then, in his treatise *Two New Sciences*, that Galileo was to prove how a moving body continues to move without being thrust by any force. The important thing is that Kepler departed this world having made a substantial contribution to the rational and mathematical understanding of it. "How can one summarise a life such as Kepler's?", asks David Park. "How many lives have shown such energy, such imagination, such sense of purpose, such ability to see facts in an entirely new way, such willingness to change one's mind?"[27]

It is tragic that after such great glory, Kepler endured years of poverty and despair. Threatened with religious persecution—which looms large in the history of the Roman Catholic Church—he was forced to abandon Prague, and wander from city to city in the midst of the Thirty Years' War. He lost his wife and children, and was forced to defend his mother against accusations of being a witch. Yet never for a moment did he lay his pen down. With his very last savings he published a monumental synthesis of astronomical data, based on the observations of Tycho Brahe. In a letter shortly before his death, he was to write: "When storms rage and we fear the shipwreck of the state, there is nothing nobler for us to do than let down the anchor of our studies into the peaceful ground of eternity."[28]

Many great scientists whose lives were marked by turbulent periods in history were to take this advice. There can be no denying that science and art, erotic love and religious faith lend human beings the strength to overcome the trials and tribulations of the surrounding world or add to their number. Perhaps because as processes, all of them originate in the deepest core of our existence.

1.9   A Messenger from the Stars

Just as Kepler sought solace in eternity, so Galileo's work will surely last for all eternity, even if he himself did obey the dictates of the Catholic Church to escape death at the stake. The persecution of Galileo appears bizarre indeed, since in his view there was no contradiction between science and Scripture. "The holy Bible and the phenomena of nature", he wrote, "proceed alike from the divine Word, the former as the dictate of the Holy Ghost and the latter as the observant executrix of God's commands."[29]

In actual fact the gap was extremely wide. With the telescope he constructed, Galileo entered what had hitherto been the sanctum of divine perfection, where its laws were exhibited. For the first time he observed myriads of stars, exceeding all known ones at least tenfold. In place of angels, he saw craters and mountains on the face of the moon. The sun had been regarded as a pure source of divine light, but he determined that it was in fact a "fiery rock", as Anaxagoras had prophetically termed it, and had unattractive spots on its surface. In contravention of divine order, he discovered that Mars too had satellites. In his treatise *The Sidereal Messenger*, published to great acclaim in 1610, Galileo writes in characteristic fashion: "I have now finished my brief account of the observations which I have thus far made with regard to the moon, the fixed stars, and the galaxy. There remains the matter that seems to me to deserve to be considered the most important in this work. That is, I should disclose and publish to the world the occasion of discovering and observing four planets never seen from the beginning of the world up to our own times, their positions, and the observations made during the last two months about their movements and their changes of magnitude."[30]

The harmonious world of celestial spheres surrounding the Earth, built by the genius of Aristotle and gladly accepted by the Church and the university establishment of the time, appeared to be collapsing. Observation and natural laws were now taking the place of manifestations of divine will and order. Reaction from the Church was only to be expected.

Galileo was born in Pisa in 1564, the year of Michelangelo's death and Shakespeare's birth. He showed considerable talent for mathematics. Despite not holding a university degree, in times very different from our own he was appointed professor of Mathematics at the University of Pisa. Shortly thereafter he moved to the much larger University of Padua, where he did the most significant part of his work in science. His reputation grew constantly; students from all over Europe flocked to his lectures, and in 1610 he earned the coveted title of First Mathematician and Philosopher to the Duke of Tuscany.

Inspired by his discoveries, he had begun to defend the Copernican heliocentric system, provoking the wrath of the Church. Cardinal Bellarmine gave him an official warning to abstain from the heretical views that claimed the Earth revolved around the sun, and Galileo seems to have obeyed for a time. Yet he returned with a vengeance in 1632, in his book *Dialogue Concerning the Two Chief World Systems*. The book was banned, he was summoned to the Inquisition and, under threat of torture, he recanted. His declaration is characteristic not only of an era, but also of a religious mentality that has not entirely disappeared even today: "But, whereas – after an injunction had been judicially intimated to me by this Holy Office to the effect that I must altogether abandon the false opinion that the Sun is the center of the world and immovable and that the Earth is not the center of the world and moves and that I must not hold, defend, or teach in any way whatsoever [...] I abjure, curse and detest the aforesaid errors and heresies and generally every other heresy, and sect whatsoever contrary to the Holy Church [...]"[31] It is worth noting the eminent scholar Stillman Drake's argument that the reason behind Galileo's trial and conviction by the Holy Office lay not in his flouting the Church, but in the enmity of contemporary philosophers![32]

Whatever the reason was, Galileo was placed under house arrest, and spent the remaining years of his life at home near Florence. There he wrote *Two New Sciences*, one of his most important works. His last discovery with the telescope was made shortly before he was struck blind, and concerned the librations of the moon. Galileo departed this world for good in 1642. The Earth did of course continue to revolve around the immovable sun, just as it has done for millions of years.

Yet Galileo's great contribution to science was not limited to the worlds revealed by his ever-improving telescope. He was a messenger from the stars, though one who never ceased to concern himself with our terrestrial world. He thus showed great interest in the phenomena of mechanics. He studied not only the parabolic trajectory of projectiles, but also the motion of bodies along inclined planes. By observing a lamp in the cathedral at Pisa, he also formulated the principle that the oscillations of a pendulum are isochronous (of equal duration), which proved highly useful in measuring time. What is more, overturning Aristotelian convictions and running contrary to common sense, Galileo proved that if air resistance is discounted, all bodies fall at the same speed, whether made of feathers or iron. He combined the insight of observation with experimental investigation and a profound knowledge of mathematics. Unbeknown to himself, he thus laid the foundations of modern experimental physics.

Given the breadth of his work and the fact that the telescope constituted a significant application of the properties of light, it may at first seem something of a paradox that light itself barely concerned Galileo. Being more of a practical than a theoretical mind, he was perhaps more interested in getting the telescope to work correctly than in how it operated. Nevertheless, his great intuition led him to perceive that like heat, light is better understood as particulate and mechanical entities. He thus drew the conclusion that when a substance is reduced to truly indivisible particles, light is produced. So light may well be some kind of body like the others; it is merely smaller, or possibly the smallest of all. Indeed, shortly before his death Galileo stated that as far as light was concerned, he himself remained in the dark. Three centuries later, having made a decisive contribution to the understanding of light, Albert Einstein was to declare much the same.

Although Galileo makes little mention of light and its properties in his oeuvre, what is plainly evident is the shift that called for it to be studied thereafter as an actually existing body with certain properties, and not a divine or spiritual substance. As an individual, Galileo also stands as a graphic illustration of the constant struggle between the spirit of science and religious dogma and persecution, particularly by the Roman Catholic Church. The fact that in 1992, i.e. close on four centuries later, Pope John Paul II was to lift the edict against Galileo, serves more to underline the importance of this struggle than to negate it. It is characteristic that even recently, the very same Church was to express its displeasure over the attempt made by modern physics to interpret the creation of the universe. As Stephen Hawking notes, in an ironic aside on a conference about problems in cosmology: "At the end of the conference the participants were granted an audience with the pope. He told us that it was okay to study the evolution of the universe after the big bang, but we should not inquire into the big bang itself because that was the moment of creation and therefore the work of God. I was glad then that he did not know the subject of the talk I had just given at the conference. I had no desire to share the fate of Galileo [...]"[33]

Yet Galileo's fate was not truly bad. Today, he is not only considered the pioneer of modern science, but folk legend also insists that he resisted the pressure exercised by the Church, uttering the celebrated phrase "And yet it moves!" immediately after his indictment. Despite this being a myth, it shows the expectations ordinary people have of their intellectual leadership. All the same, scientists are no exception to the fact that in real life compromises are more common, and acts of constancy and heroism rare.

1.10   The Rational Reconstruction of Knowledge

At any rate, Galileo's fate was at least escaped by the third leading philosopher of the time—René Descartes. He was intending to publish *The World*, a treatise in which he supported the Copernican heliocentric system, but withdrew it as soon as word reached him that the Holy Inquisition was persecuting Galileo for the same reason. An important work, it was published long after his death, and even then with a revised content.

Descartes' life was highly idiosyncratic. Indeed, his intellectual career has little in common with the medieval tradition that hallmarked Kepler and Galileo. He was born in a small town in France in 1596 and gained a broad education, firstly at a Jesuit college and then at Law School in Poitiers. Keen to travel and think, in a somewhat unorthodox way of seeking reflection he joined the army of a contemporary prince. Thus, as a non-combatant solider, from 1618 he spent an entire decade touring many European countries. It was then that he began to work on mathematics, conversing with prominent figures in the world of science as well as with mystics of all kinds. He eventually settled down in Holland, possibly because the Church authorities intervened less there, though Descartes remained a Catholic for his entire life. It was at this time that he published the greater part of his multi-faceted and innovative work. In the last year of his life he was to move to Sweden, where he had been invited to teach philosophy to the queen. This must have been a most onerous duty for Descartes, since he was used to spending all day in bed reading and thinking. Whatever the case may be, he breathed his last in Sweden in 1650, bequeathing posterity a unique oeuvre in philosophy, mathematics and optics.

Although Descartes himself preaches rational thinking, it is interesting to note that he conceived of his philosophical system in the classic visionary manner. He recounts how one night, in the course of his endless and rather aimless wanderings as a soldier-philosopher, he had a vivid dream—in three parts, no less. This included nightmares, thunderbolts and inexplicable flashes of light, ending in a line of Ancient Greek that revealed the path he should follow. As interpreted by Descartes, the dream pointed out the need for a new method of seeking the truth.

Descartes attempted to accomplish this mission a few years later, by the publication in 1637 of his classic work *Discourse on Method*. The preamble refers in characteristic fashion to "The Method of Rightly Conducting the Reason and Seeking for Truth in the Sciences", and the work contains sections on optics, meteorology and geometry. As in his later works, Descartes bases the reacquisition of knowledge on systematic doubt. His famed saying *Cogito, ergo sum*—"I think therefore I am"—was the cornerstone of his philosophical system. Yet that system was built with strict rationalism and followed axiomatic methods of mathematics as taught by Euclid.

From the simplest atom to a complex organism, Descartes' universe is a mechanism that works by mathematical logic. "The rules of nature are the rules of mechanics," he declares; a view which was to reach its apotheosis in Newton a few years later. Descartes also believed in "dualism", i.e. in the distinction between spirit and matter, while to his mind God ever remained a supreme, perfect Being.

It should be noted that the influence exerted by Descartes' philosophy had its ups and downs. But the same did not occur with his contribution to mathematics, which was lasting and without contest. Thus the "Cartesian" system of co-ordinates established by him led to many problems in geometry being solved via the algebraic method. Descartes also contributed to shaping astronomy, and studied anatomy and animal behaviour.

The nature of light and the operation of vision were of particular concern to Descartes. Both in *Dioptrics* and at other points in his diverse oeuvre, he applies his axiomatic method. He concludes that light is a kind of pressure—not motion, at any rate—in the densely compressed medium that is only apparently empty. This pressure is transmitted instantly from particle to particle, eventually reaching the eye. Indeed, he compares this mechanism with a blind person's stick—as soon as it touches an object, the feel of it instantly reaches the hand, without anything moving inside the stick or travelling towards the soul.

Thus according to Descartes, the nature of light does not resemble that of a projectile or fluid; it is more a tendency towards motion, and is transmitted at infinite speed via the material substance that fills all of space. As the historian of science A. Sabra observes: "The Cartesian theory was the first clearly to assert that light itself was nothing but a mechanical property of the luminous object and of the transmitting medium. It is for this reason that we may regard Descartes' theory of light as the legitimate starting point of modern physical optics."[34]

With characteristic ease, Descartes does of course sidestep certain substantial difficulties, declaring that light is a tendency or inclination towards motion, but does not entail true motion. Nevertheless, in this vague view lies the seed of the wave nature of light, which was later to overthrow Newton's particulate theory. But it was also to undermine Descartes' dream itself, which aspired to construct a purely mechanical universe.

Descartes' studies on vision appear more down to earth. In *Dioptrics* he gives a detailed description of the course taken by light rays, and elucidates how the pupil, the lens and the retina work. He also makes particular reference to lenses, and makes a somewhat unsuccessful attempt to contribute to improving them and enhancing the sharpness of the telescope.

It is thus obvious that Descartes' *Dioptrics* does not shine for its originality or practicality. His philosophical work, too, is often criticised for its contradictory or arbitrary reasoning. Yet the value of his philosophy lies elsewhere—in the fact that he avoided the pedantry and vague generalisations of his time, and clearly highlighted his aim and intentions. The origin of truth is not exogenous; on the contrary, the brain gains the truth from what it observes and concludes. Descartes was the first person after Aristotle to try and organise knowledge into an integrated system, and to persuade others of his method. He is thus rightly regarded as one of the great figures of western thought. "Since the thirteenth century," the eminent historian Fernand Braudel comments, "Western science has lived with only three general explanations or world systems: that of Aristotle, which although of ancient lineage entered the interpretations and speculations of the West in the thirteenth century; that of Descartes and Newton, which founded classical science and which is an original Western creation except for its decisive borrowings from the works of Archimedes; and finally the relativity theory of Albert Einstein, announced in 1905, which inaugurated contemporary science."[35]

It is now time to leave the great progress that led to the scientific revolution, with a brief mention of another eccentric figure from the period: the French mathematician Pierre Fermat. Like Descartes he studied Law, and even became a parliamentary advisor. But mathematics always remained his true passion, and the most important part of his work—usually results without proofs—was set out in letters to his friends. His correspondence with Pascal even laid the foundations of probability theory; by Newton's admission, the first elements of differential calculus are to be found in Fermat.

As the most familiar manifestations of a light beam, the phenomena of reflection and refraction also concerned Fermat. The simple law of reflection had already been formulated by Euclid. Refraction, on the other hand, is obviously more complex. So after a first, passable approach by Ptolemy, it took several centuries until the precise law was formulated in 1621. The credit for discovering it goes to the Dutch mathematician and geodesist Willebrord Snell, after whom it is still named in school textbooks. Snell's Law links the angles of incidence and refraction of a light ray passing through a material medium, thus proving that the relationship between them is a property of the material.

Shortly after the law was formulated, Fermat offered a radical interpretation of the particular way light behaves. According to this interpretation, it quite simply and invariably selects the route that takes the least time. This holds true both when light is reflected off a surface, as already observed by Hero of Alexandria, and when it is refracted in a material medium. Fermat's principle thus unifies both phenomena, revealing the wonderful economy of nature. For light, like hurried commuters driving in a modern city, the criterion when moving from one place to another is never distance, but the time a route takes. As Fermat proved in mathematics, the law of refraction is a simple consequence of that principle. Indeed, while in the course of formulating it, Fermat had corresponded with Descartes, who ignored his observations despite having invited discussion!

Nevertheless, Fermat's principle turned out to be of much greater significance. A wide range of phenomena in optics and hydromechanics, quantum theory or mechanics can be accounted for on the basis of similar reasoning, which obey the more general "principle of least action". This is just one manifestation of the marvellous economy concealed in many of nature's laws. In fact, the principle of least action was to reach its apotheosis when Richard Feynman, one of the greatest physicists of modern times, applied an ingenious variation on it to studying interactions between light and matter.

Be that as it may, it should be stressed that the dawn of the scientific era was not hallmarked by the dazzling presence of Kepler, Galileo and Descartes alone. Nor did Ancient Greek thought comprise exclusively of Democritus and Empedocles, Plato or Aristotle. Every era has its outstanding figures, whose names are later recorded in school textbooks and in collective memory. But in parallel with them there have always been important thinkers who, for one reason or another exerted less influence or have remained in the half-light. The fact that no mention is made of them in the previous pages, and in those to come, does not mean that their contribution has been ignored. Science in particular relies on the collaboration of many. Each person has their own small contribution to add, or their own hue to colour the enormous structure that is science. Great figures in science often appear to be solitary, yet they never cease to rely on the outstretched arms and familiar voices of others.

1.11   The Luminous Parallels Converge

We have seen how, over the course of roughly twenty centuries, human preoccupation with light has taken two parallel paths (Fig. 1.3). One has its roots in Ancient Greece and, in a somewhat elastic sense of the term, can be called scientific; it attempts to trace the properties of light and asks questions about its nature.

The second path is defined by the experience of light in its religious or mystical incarnation. This experience, which marks out the ascent of man from the world of darkness and ignorance into a spiritual world, is encountered in its various forms in many religions. In Platonic and Neoplatonic texts, in Indian philosophy as well as in the hesychastic experiences of the ascetics, in the Fathers of the Church and the great Christian thinkers, light diffuses powerfully, leading to revelation and knowledge. Historian of religions Mircea Eliade calls this "internal light", summing up as follows: "Here lies the paradox: the meaning of Light is, on the one hand, ultimately a personal discovery; and, on the other hand, each man discovers what he was spiritually and culturally prepared to discover. Yet there remains this fact which seems to us fundamental: whatever will be the subsequent ideological integration, a meeting with the Light produces a break in the subject's existence, revealing to him – or making clearer than before – the world of the Spirit, of holiness and freedom..."[36]

The truth that light is a personal discovery was something the author of this book also knew from his own, solitary point of view. As he wrestled with light for years, he had the feeling that that woman—who had been accompanying him for centuries now, around the seas and islands of the world—knew how to travel along infinite pathways of light, experiencing light as a redemptive, everlasting discovery.

From the sixteenth century onwards, the scientific revolution, which slowly but surely liberated man from religious dogmas, was to place particular emphasis on the investigation of light. Up until then, cases where the two paths of science and internal light intersected were frequent and unavoidable. Science then moved ahead self-confidently if somewhat arrogantly, also resting on the irrefutable authority of experimental methods.

Fig. 1.3 An archangel explaining the physical nature of the universe, etching by James Barry (1795). Upper register (left to right): Francis Bacon, Copernicus, Galileo, Newton, the archangel, an angel; lower register: Thales of Miletus, Descartes, Archimedes, Grosseteste, Roger Bacon (The Wellcome Collection/CC PD)

Nevertheless, the scientific path to the understanding of light holds both major achievements and surprises. Light has proved recalcitrant and wondrous by nature; yet at the same time it is a fundamental entity in physics, and its contribution to investigating the universe has been truly unique. The interesting thing is that as scientific knowledge reached its apogee in the twentieth century, the understanding of light left room for the unexpected and the timeless, the wondrous and the uninterpreted. An internal light seems once again to be rising. And while this time it does not reflect religious origins, it is of indisputable existential quality, linking human beings to a universe that surrounds and transcends them.

2

Light's Virtues and Vices

2.1   A Scientific Prelude

Science follows an established methodology when tackling a problem. If, for instance, the problem concerns a natural entity—the nucleus of an atom, a star, the molecule of heredity—science strives to gather as much information as possible on that entity, and to record its properties and behaviour. Once at least a partial record of that behaviour has been created, an attempt is made to interpret it on the basis of a more general scientific theory, which normally takes years to formulate. Thus, it only became possible to understand the chemical properties of molecules once quantum theory had been developed; the true nature of time emerged from the theory of relativity; and the correct answer to the simple question "Why do the stars shine?" ("Because of nuclear processes") was only given a few decades ago.

The breadth of phenomena or experimental results that a theory encompasses are the judge of whether it stands the test of time and becomes definitively established. Otherwise, the theory has to be augmented, possibly revised and in some cases refuted in its entirety. When Newton interpreted existing experimental data on the basis of the hypothesis that light consisted of material particles, there was a rational and theoretical consistency to his interpretation. But soon doubt began to be cast on this, as newer results showed the wave nature of light.

In any event, science too is an integral part of culture; when we talk of culture and simply mean literature, artworks or music, we are taking away a large part of its dynamics. The theory of relativity, for instance, has its own rare aesthetic, and opens up new dimensions to human beings just like any

G. Grammatikakis, *The Autobiography of Light*, https://doi.org/10.1007/978-3-031-56917-3_2

major artwork. Furthermore, it has assisted us in understanding why the sun shines. For similar reasons, the discovery of the helical structure in DNA molecules or differential calculus in mathematics are cultural events, in any sense of the word. Internal beauty and timelessness are just as characteristic of scientific truths as they are of great artworks.

Interaction and the constant interplay between experiment and theory, predictions and refutations are what make for both the splendour and the pitfall of true science. Splendour, because experiments provide a true testing ground, a safe criterion for truth and how far it goes. The experimental process is not about a set of outlandish instruments joined together by tubes and wires, as common belief would have it. An experiment of the type that makes its mark on the history of science calls for an admirable combination of ingenuity, technical skill and knowledge.

Those same features also make for the perilous pitfalls of experimental science, since there is often the tendency to force data to support a current or widespread theoretical view. "First get your facts," Mark Twain jokingly advises, "then you can distort them at your leisure." One characteristic instance is the aether; even up to the beginning of the previous century, its existence appeared necessary in comprehending the behaviour and propagation of light. Nevertheless, the genius of Einstein interpreted the experimental data in a radical upset of perceptions, which not only did away with the aether, but simultaneously paved the way for a true understanding of space and time.

Yet another substantial pitfall lurks in the mandatory verification of scientific ideas by experiment (repeated, different experiments, no less), which lends lustre and certitude to the natural sciences. Many of the major scientific hypotheses are often extremely difficult if not impossible to confirm by experiment. Contemporary cosmological theories are one shining example. However optimistic physicists may be—and they often do go overboard—it is impossible to replicate the Big Bang in terrestrial laboratories. So the truth of cosmological perceptions is judged by how much they fit in with present-day observations, and by their internal consistency. Besides, in cases where a theory aims at substance and universality rather than partiality, part failures are not of great consequence. To this day, Democritus' theory of the atom remains correct in conception, despite the fact that the true nature of atoms was only revealed by the twentieth century. Thus though the uncritical acceptance of insight and intuition often proves dangerous, their importance in science is beyond doubt—and this is another area where science borders with art.

In any case, when verification by experiment is not feasible, and is replaced by divergent thinking or intuition, branches of learning different from the natural sciences also take similar paths. Archaeology, for example, slowly reconstructs the past on the basis of written sources and excavation finds. The thrill felt by an archaeologist when an unearthed building or tomb adds harmony to the song of the past is no different from that of a scientist when a theoretical prediction is confirmed by experiment. Besides, only ignorance and hysterical figures in the Church still maintain opposition to Darwin's theory of evolution, but the original "experiment" is impossible to repeat since, apart from anything else, it lasted millions of years! Darwin himself and his successors thus relied on observations of the living world and painstaking data collection. With some fine tuning, the initial grand conception was confirmed.

It is true that the impressive progress made by the often deceptively termed "hard" natural sciences, mainly over recent decades, has led to their influence swamping the so-called humanities. For example, sociology and psychology have both attempted to imitate hard science methodology, though often with trite results. Philosophy, on the other hand, often forgetting the fundamental questions it poses, is rushing headlong in an attempt to understand the latest developments in physics.

Yet the fact still remains that the unification of knowledge remains a noble vision. To realise it, leading evolutionary biologist Edward O. Wilson envisions a "consilience" or joint leap forward by the natural sciences together with the social sciences and the humanities. "The Labyrinth," he writes, "its likely origin a prehistoric conflict between Crete and Attica, is a fitting mythic image of the uncharted material world in which humanity was born and which it forever struggles to understand. Consilience among the branches of learning is the Ariadne's thread needed to traverse it. Theseus is humanity, the Minotaur our own dangerous irrationality."[1]

That being said, it is not the aim of the present chapter, or indeed of this book, to go deeper into epistemological questions or their philosophical ramifications. This is and remains a book about light. And perceptive readers will already be suspecting that light is in parallel a pretext, allowing the author to talk about other matters of his soul and the world. But if the above thoughts on the limits of experimentation and theory seem unduly long-winded, then the reason is this: what characterises the investigation of light more than anything else is the constant interplay between theoretical ideas and experiment; and the quest for its true nature forges ahead through triumphs and falsifications, amendments or quantum leaps. So the history of light has all the fascinating feel of a novel, but its protagonist seems to be constantly changing faces, flashing the reader a sardonic smile.

2.2   Light's Manifest and Hidden Graces

We know about some of light's obvious properties and features of its behaviour from experience. For instance, the fact that it travels in a straight line. Car headlights cannot possibly shine out of the direct line of sight around a hairpin bend, however useful that would be. Nor can we admire a beautiful woman once she has turned the corner. But if light behaved like sound—which can bend around objects—then the world would look different. It would resemble an endless symphony of colours, where objects were blurred and ethereal. Night would not exist.

The linear propagation of light has one key consequence in everyday life. When an opaque object blocks the path of light rays, the shadow created varies greatly in shape and intensity. From the shade of a tree to eclipses of the sun and moon, from shadow theatre to the shadow that follows us, shadows are part and parcel of our world and lives.

It comes as no surprise, then, that shade and its symbolism are used in the patchwork of our lives. The Greek national anthem sings of liberation from "the shadow of tyranny", while "she lives in his shadow" is said of a woman whose fate it is to live with some (usually overweening) personality. There is also an adjective in Greek that roughly translates as "light-shadowed", used for daydreamers who are out of touch with reality—an attribute which the cynicism of our times tends to regard as engaging.

Be that as it may, eclipses of the celestial bodies are one breath-taking phenomenon caused by the existence of shadows. Solar eclipses are due to the moon intervening between the sun and the Earth, and were regarded by ancient peoples as important omens. An eclipse of the moon, on the other hand, is caused by the shadow of the Earth created by sunlight. Numerous eclipses are cited in the astronomical records of every people, and often also in their literary and historical texts. Typical of this is the poetic manner in which Pindar comments on a solar eclipse: "Beam of the Sun, what have you contrived, far-seeing one, O mother of eyesight, supreme star, by being hidden in daytime?"[2]

Let us now return to prosaic yet light-filled reality. How does light behave when it encounters surfaces or bodies that are at times transparent and at times block its path? As careful observation reveals, light reflects off smooth surfaces: indeed, a light beam leaves a surface at the same angle it has encountered it. That's how we see our face in the mirror, and also see a car overtaking us in our wing mirror. It is important to stress that "see" always means that light rays leave or bounce off the object and reach our eyes.

A mirror is smooth, of course, but a wall or city street is not. Such dimpled surfaces reflect irregularly—in other words, they diffuse light. Diffusion is what makes objects around us visible. It is why day breaks long before the sun comes up, as sunlight is diffused by particles in the atmosphere. So it comes as no surprise that on the moon, where there is no atmosphere, an astronaut's shadow looks stunted and very dark.

Several of our paradoxical experiences are due to refraction, another of light's well-known attributes when it travels to the edge of a transparent medium. A spoon dipped halfway into a glass of water appears broken in two, while the world viewed through a bottle appears unfamiliar and distorted. Light rays change course on refraction, obeying a simple mathematical formula that has to do with the nature of the transparent material. Even the sun's rays are constantly refracted as they pass from higher, thinner layers in the atmosphere to layers of air near the Earth. This atmospheric refraction causes the sun and stars to appear higher than they are in reality. What is more, the apparent twinkling of distant stars is due to the constant refraction of light.

Close observation, then, reveals light's many graces. Yet it also adds to the suspicion that its behaviour must be governed by some fundamental law. This law has to interpret two contradictory observations: the linear propagation of light when travelling in uniform media, and its refraction when passing from one optical medium to another. As we have seen, the law in question was formulated by Fermat, and is wonderfully simple: light follows the path that takes the least time. Thus in air or any other transparent medium it takes a linear path, since a straight line is the shortest route for any motion at a steady speed. In differing substances, however, the shortest route is no longer linear. Light travels at a different speed in each medium, and so is forced to refract at the point of contact precisely in order to find the shortest path. Readers may well be wondering how light knows in advance what the shortest path is; and at the end of the day, why it should be interested in discovering that path. If we extrapolate, it becomes obvious that the answer has to do with the paradoxical nature of light itself.

Sidestepping the question, to say nothing of the answer, it is worthwhile returning to the behaviour of light. We have seen that some of its features, such as refraction and reflection, are even open to simple observation. But other elements of its behaviour—its hidden graces—are not easily deductible in the same direct way. Indeed, since Newton's authority cast its long shadow over every opposing view in around the eighteenth century, when the debate over the true nature of light had grown heated, it took delicate and persistent experiments to bolster heretical ideas.

We saw earlier how reflection and refraction force light to diverge from its regular, linear course. But there is also one further way it diverges: by diffraction. If light passes through a narrow slit, made for instance by a razor in cardboard, the shadow behind the cardboard does not have a clearly defined border between lit and shaded areas. The light spreads like a fan, gradually decreasing in intensity until it becomes totally dark. As we say, it has been diffracted.

Diffraction is not limited to narrow slits or openings, but can be observed in any kind of shadow. Close up, even the sharpest of shadows is slightly blurred at its edges. If you put your fingers together and look through them at a light source, you will see black lines and shapes caused by diffraction. In fact, where light diffracts at the edge of an opaque object, thin bars of colour appear to intrude between the light and dark areas. It is as if light is wrestling with darkness, and colour is born of the struggle.

In all of light's behaviour, the truth is that diffraction appears to be an innocent and somewhat trivial aspect. Nevertheless, it was one of the most compelling indications of light's wave nature. Something similar can be observed among people—a subtle gesture or look often leads to a profound and carefully hidden side of their personality.

2.3   The Wonderful World of Colours

Colours are light's greatest grace, artfully concealed in its white nature. This is a truth hard to take in, since colour appears to be an integral attribute of every object, a kind of property it owns. Concern over whether colour is due to light or objects themselves goes back to the Ancient Greeks. It took many centuries before the correct answer, which is "to both", gained currency. White light really does consist of many colours, but a given body only reflects a few of them, while absorbing others. So the colour we see is a synthesis due both to the attributes of white light and to those of the object itself.

Deep down, however, colours are something more than light, and something more than the behaviour of an illuminated object. Colours are also "perception"—in other words, a complex creation of the eyes and brain. As shown by the scientific study of colour, beginning in the nineteenth century and continuing apace to this day, it is not solely a problem of physics, nor solely one of physiology. It lies in the area where the two disciplines overlap, bordering on the psychology of perception.

Interpretations aside, the variety of colours in nature is truly unlimited: sea pebbles and sunsets, birds and multicoloured flowers, a rainbow or the colours of a butterfly. Indeed, humans imitate and add to this variety with their own colour compositions. Having learnt the technique of colours and their shades, they make multicoloured clothing and jewellery, exquisite carpets and paintings.

It is thus impossible to imagine aesthetics or the functioning of everyday life without the feast of colours and the quality they lend to things and life. Note that to a great extent, insects are even directed by the colours of flowers; so it's curious that while the presence of colours has been heavily felt since ancient times, the first indications of what they really were did not come until Newton's experiments, as late as 1666. As he himself describes: "I procured me a triangular glass prism, to try therewith the celebrated phenomena of colours: And in order thereto having darkened my chamber, and made a small hole in my window-shuts, to let in a convenient quantity of the sun's light, I placed my prism at its entrance, that it might thereby be refracted to the opposite wall. It was at first a very pleasing divertisement to view the vivid and intense colours produced thereby; [...] I have often with admiration beheld that all the colours of the prism being made to converge, and thereby to be again mixed as they were in the light before it was incident upon the prism, reproduced light, entirely and perfectly white, and not at all sensibly differing from a direct light of the sun..."[3]

Though Newton's experiments now seem simplistic, they ended in the significant finding that white light is composite. A glass prism analyses light into the colours it consists of because, depending on their colour, light rays refract to differing degrees. On refracting, red bends less than all other colours, and violet more. Orange, yellow, green, blue and indigo extend between these two extremes, comprising the spectrum of white light. Even an ordinary sunset showcases the richness of the colour spectrum, and the never-ending tricks light plays with particles in the atmosphere. The sun takes on a bright orange or red colour as its sets. Its light is reflected in the clouds, colouring great swathes of sky. Until dusk and then night fall gently, the colours constantly change and darken. At beauty spots such as Sounio or Santorini, the sunset acquires a bewitching beauty, not coincidentally associated with romantic moods.

So colour leads a paradoxical existence: it is contained in light, which appears colourless to the eye. In reality there are no colours in the rainbow or a painting. We do not inhabit a world of coloured objects, but one in which the variety of texture and surfaces allows for infinite colour combinations. In his *Cosmicomics*, Italo Calvino experiences a colourless grey world, where like a colourless bolt of lightning he encounters Ayl, a female apparition. At some point, however, the lack of colour disappears: "All around, the world poured out colours, constantly new, pink clouds gathered in violet cumuli which unleashed gilded lightning; after the storms long rainbows announced hues that still hadn't been seen, in all possible combinations. And chlorophyll was already beginning its progress; mosses and ferns grew green in the valleys where torrents ran. This was finally the setting worthy of Ayl's beauty; but she wasn't there! And without her all this varicoloured sumptuousness seemed useless to me, wasted."[4] It is a fact that the beauty of the world often seems equal to that of an Ayl. On the other hand, how often is it not that an Ayl is all it takes to lend beauty to the world?

Nowadays, the fact that colours are one of white light's invisible graces seems a self-evident truth. Yet how do colours differ from each other in reality, apart from the different sense they create? The answer to that pertinent question is simple. Each colour differs from others in one attribute alone: its wavelength. A colour's frequency or wavelength are related to the wave nature of light. When the white light of the sun falls on an object, only one or a few of its colour components are reflected, depending on their frequencies. They are the ones which lend the object its characteristic colour, while the remainder are absorbed by matter. So a rose looks red because only the red component of white light—i.e. the wavelength corresponding to red—is reflected off the flower and reaches our eyes. The other colours are absorbed. By contrast, the wall of a Greek island house is white because it reflects all colours, which are reconstituted anew in white.

Even the blue of the sky can be interpreted in a similar manner. As sunlight travels towards the Earth, it is reflected off gas particles and dust in the atmosphere. But the nature of the atmosphere is such that it gives preference to reflecting blue, or the blue component of sunlight. If the Earth had no atmosphere, then things would look entirely different. One of the astronauts on the Apollo 17 Mission used the following words to describe his experience of the moon, which has no atmosphere: "I had tried to anticipate what it would be like for many years. But there was no way to anticipate standing in the valley of Taurus-Littrow, seeing this brilliantly illuminated landscape with a brighter sun than anyone had ever stood in before, with a blacker than black sky, and the mountains rising on either side."[5] It goes without saying that the sky on the moon is dark, since the only light to reach any eyes there—terrestrial eyes alone, as far as we know!—comes directly from the sun.

So light is the one and only source of colours. Its interplay with matter is the first filter determining which of its colour components are to predominate and which are to be neutralised. Thereafter the complex operation of vision intervenes, so as to create the final "sense" of each colour. Besides, it is well known that the colour of objects greatly depends on the composition of the colour illuminating them—a truth well known to those coquettish ladies who insist on checking the colour of cloth outside the draper's, in natural light. Indeed, illuminating an object with monochromatic light such as a laser beam often produces unexpected results.

So even with the naked eye, nature offers a magical, inexhaustible spectacle of light and its manifestations. And bearing in mind that the boundaries between colours are not clear-cut, that some colours are composed in others and that refraction and diffusion are ubiquitous, the infinite variety of colours in our world comes as no surprise. Light does not simply reveal the visible world, but to a great extent also creates it.

2.4   Chromatic and Philosophical Acrobatics

As experience teaches, the sun and the countless changes in atmospheric conditions have an immediate effect on the colours of the sun and the Earth. Optical phenomena vary from shadows and reflections to dawn and dusk; and from deceptive mirages to the games light plays with water. Ice crystals, which differ in shape and form, create worlds of unparalleled beauty via reflection and refraction: the sparkling and shimmering of snow, wonderful halos and coloured rings around the sun and in the clouds.

One phenomenon familiar to us is how the sky brightens and takes on a deep blue when heavy rain follows a dry spell. And that is because the dust settles, limiting the diffusion of light. Similarly, the reflection of light from the sky lends the sea its characteristic blueness. Near the beach, however, the fine sand suspended in the water diffuses light beams of a longer wavelength, and this gives us the light green colour of sea water. Where there is a great deal of seaweed, the sea sometimes appears purple. This unusual colour is due to the red light reflected by seaweed mixing in with the blue of the sky—it is Homer's "wine dark sea", which led to an incomparable line describing Crete:

> There is a land called Crete, in the midst of the wine-dark sea, a fair, rich land, begirt with water[6]

As the theory of relativity will show, only light has gained eternity. Nevertheless, Greek light appears to possess something absolute and transcendental, which mutates the evanescence of the world into an element of eternity. "So much light that even the naked line is immortalised,"[7] writes Odysseas Elytis. In the Greek landscape, the graces of light and its colours and blazes are gloriously evocative.

On another level, one phantasmagoria of light familiar to us all is the rainbow. It often appears when the winter sun plays hide-and-seek with the clouds after rain. In Homeric times, the rainbow was regarded as a sign from Cloud-Gathering Zeus, presaging war or rain, while in biblical tradition its appearance served as divine confirmation that the Flood would never reoccur. One of the many folk beliefs holds that a crock of gold is to be found at the end of a rainbow, where it meets the ground. But all effort is in vain: however much you search, you will never find the end of a rainbow, since it moves along with the person looking at it. And that is because rainbows are quite simply due to the reflection and refraction of sunlight when its path crosses raindrops. The colours in the sunbeams are then forced to scatter; and if conditions are right, then a rainbow unfolds majestically in the sky, like a row of concentric coloured arches. Of course, neither a rainbow's appearance nor its shape are simple phenomena, depending as they do on the position of the raindrops in the sky and the diverse routes taken by the sunbeams that are successively reflected or refracted. In fact, it is not uncommon for a secondary rainbow to appear, dimmer than the first and with its colours in reverse order.

Newton gives a detailed description of the rainbow as the consequence of light being refracted through raindrops in his *Opticks* (Fig. 2.1), though as we have seen, an initial explanation for the phenomenon is to be found in the work of Aristotle, and we owe our definitive understanding of it to Descartes. Indeed, the rationalist philosopher grabs the opportunity to reiterate his conviction that logic and observation suffice to explain physical phenomena.

On the other hand, the English Romantic poet John Keats believed that reducing the rainbow to the colours of the spectrum destroyed all of its poetry. This is a common misunderstanding regarding the role of science, expressed in extreme form by one of the most brilliant intellects in recent centuries: Johann Wolfgang von Goethe. A poet, essayist, dramatist and politician, Goethe is regarded as the founder of modern German literature. Yet at the same time he took an active interest in science, particularly in anatomy, biology and physics. Since his scientific convictions were typified by a combination of highly acute observation and poetic insight, he rapidly grew disenchanted with the dry, static taxonomy of plants developed by Linnaeus, which left no room for the imagination and beauty of a flower.

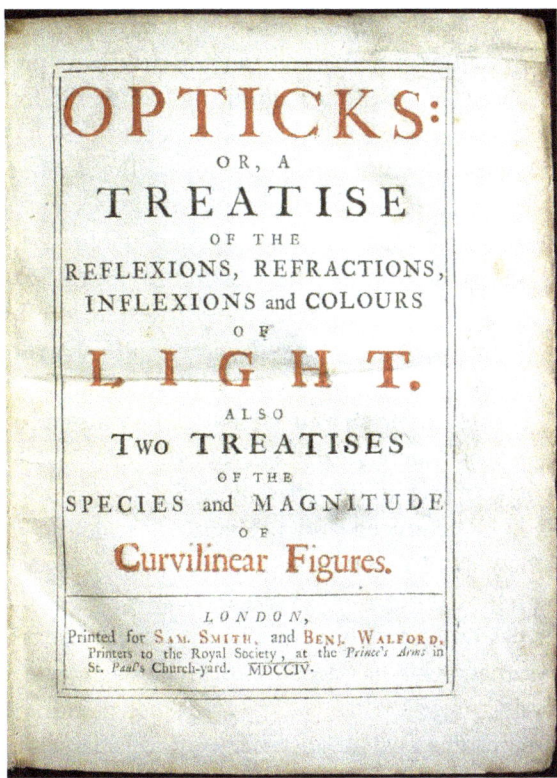

Fig. 2.1 The title page of Isaac Newton's Opticks (1704) (Wikipedia Commons/PD)

Goethe was not long in turning for consolation to light and its everlasting manifestations. On a journey to Italy he had already been bewitched by the brilliance of the landscape and the clarity of colours. Wishing to approach them scientifically, he had recourse to Newton's *Opticks*. Here his sense of disillusionment was once again keen, for he felt that the Newtonian analysis of light fragmented the whole. So in 1810 he published his *Theory of Colours*, a treatise in which he rounded on Newton, treating colours with poetic verve as *Taten und Leiden* ("actions and passions") of light. Goethe believed that phenomena should be observed free of artificial constructs, under the open sky and with the senses fully functioning and alert. As he concluded: "Light is the simplest, most elementary, most homogeneous thing we know. It is not compounded."[8] It goes without saying that Newton was vindicated by science and experimental verification. Indeed, Goethe has often been criticised for his pseudoscientific theories, though this does nothing to detract from his towering intellectual presence.

In any event, the philosophical meaning contained by the world of colours is no easy matter. It has thus quite rightly also preoccupied great thinkers other than Goethe. One significant contribution was made by Ludwig Wittgenstein, a leading—and largely inaccessible—twentieth century philosopher. References to colour are to be found throughout his work, and his last notes were published under the characteristic title *Remarks on Colours*. As he writes: "When we're asked, 'What do "red", "blue", "black", "white", mean?' we can, of course, immediately point to things which have these colours, - but that's all we can do: our ability to explain their meaning goes no further. For the rest, either we have no idea at all, or a very rough and to some extent false one."[9] Of course, it goes without saying that the properties of light and the chromatic behaviour we have just outlined—its visible and invisible graces—do not merely exert charm on poetry or philosophy. They are also of great importance in understanding light's basic mechanism. How, one wonders, is light produced and propagated, and how does it behave? In other words, how can its seemingly ever-evasive, ever-changing nature be understood?

As the author has reflected from time to time, light has something of the charm of that woman who has accompanied his wanderings for centuries now. Like light, her presence has always given rise to questions and admiration, leaving one to guess at some of her cryptic facial expressions; yet her soul itself has never seemed easy to uncover. Possibly forever, she has remained inaccessible to the superficiality of glances and observations.

2.5   A Speed Different from Others

We have seen how certain of light's properties can easily be deduced by close observation. In that way we know that light is reflected, refracted and can be analysed into many colours. But observation is of no substantial assistance in investigating the profounder question of the speed at which light travels.

To some this question may sound meaningless. Light is everywhere, and doesn't appear to "speed" like a runner or, for argument's sake, an aeroplane. Yet we get quite the opposite sense when we turn on a torch, and it instantaneously lights up a wall opposite us. It is as if the light were being catapulted from its source at unimaginable speed. Besides, we are all familiar with one phenomenon characteristic of thunderstorms: when lightning strikes, we see the flash before we hear the thunderclap. In other words, the light from the atmosphere where lightning is formed by electrical discharges reaches us much faster than sound, and so must be travelling at greater speed.

By human standards, the speed of light really is mind-bogglingly colossal. As any school textbook will tell you, it is in the region of 300,000 km per second! In other words, in a mere second light goes more than seven times round the world. Inquisitive readers need only compare the speed of light with that of a car, an aeroplane, or even of sound itself. In air, it is almost a million times greater than the speed of sound.

Given that, it's hardly surprising that measuring the speed of light took centuries of reflection, debate and ingenious experiments. The suspicion that it was incredibly fast if not infinite was first aired in Ancient Greece, and was linked to its particulate nature. As we have seen, such ideas were vehemently opposed by Aristotle, who insisted that light was a potential "state" of the medium, so the notion of it being propagated was meaningless.

Much later, the view that light travelled at infinite speed was also supported by St. Augustine, who, we should note, linked God to light. And as was theologically necessary, God had to be ever-present everywhere.

Descartes was another figure to argue passionately for the instantaneous propagation of light, since it fitted in with his more general conception of the mechanism of sight. He even wrote that if his hypothesis was disproved, he would renounce all of his philosophy. Luckily enough, his magisterial oeuvre avoided that accursed fate, as he breathed his last long before 1675. For it was then that by observing the moons of Jupiter and their periodic eclipses, the Danish astronomer Rømer proved that light must indeed travel at an exceptionally high speed, yet one that was finite.

Over the centuries following Rømer's first measurement, which established the correct framework for its magnitude, the speed of light stood as a major challenge for experimental physics. Various methods and techniques were employed: terrestrial and astronomical ones, revolving mirrors and interferometers, and recently even laser beams. The accuracy of measurements constantly improved. For instance, the eminent experimental physicist Michelson spent the last twenty years of his life working on them. In 1931 he even constructed an evacuated cast-iron pipe a mile long, where light travelled freely back and forth and was recorded by sensitive devices. The measurement results gave an even more precise value for the speed of light, but they found Michelson ill in bed, two days before he departed this life - obviously a happy man!

The speed of light is now taken for granted, and physicists no longer concern themselves with calculating it. As was concluded by the best measurement up to 1983, which was agreed on as the reference value, light travels in a vacuum at 299,792,458 m per second.

The magnitude of the speed of light is without doubt awe-inspiring. Even the high speeds we are familiar with from experience or learning cannot compare with this tremendous velocity: the speed of a rocket does not exceed a few dozen kilometres per second, while the orbital speed of the Earth around the sun is 30 km per second.

Yet the speed of light is not simply colossal. As Einstein proved, it is not only the greatest speed known, but the greatest that can exist in nature. Nothing is permitted to exceed it. In fact, it remains the same even if the source emitting a bundle of light is moving with it.

Of course, light slows down when travelling in transparent materials such as glass or water; its speed in water falls by a quarter, and in diamond even more so. We should in fact note that in water, charged particles such as electrons are capable of gaining high speeds even surpassing that of light. This produces so-called Cherenkov radiation—named after the Russian physicist who discovered it—similar to the loud sonic boom heard when an aeroplane breaks the speed of sound.

The above facts do not influence the special characteristics of the speed of light. When it passes through transparent matter and emerges unaltered into the familiar, boundless void, light finds itself again, and regains its constant speed, which is also the fastest in nature. So no matter how one looks at it, the speed of light differs from that of a projectile, or an electron in a conductor. More than merely a number, it appears to be a fundamental key to understanding the world, even if the locksmith remains unknown.

In fact, the high value of the speed of light renders it suitable for measuring the colossal distances in the universe. The distance light travels in one year, moving constantly at 300,000 km per second, is somewhat misleadingly called a "light year"—a distance approximately equal to ten thousand billion kilometres! By this unit of distance, Sirius, the brightest star in the winter sky, is 8 light years away from us. On the other hand, Andromeda, which is the closest galaxy to our own, is roughly two and a half million light years away. Given that no spaceship is capable of breaking the speed of light, that fact renders any visit by humankind unlikely, though it has not prevented science fiction novels and films from regarding Andromeda as a permanent battleground between extraterrestrials and earthling space warlords.

Yet the true significance of the speed of light does not lie in its incredible magnitude. As has been revealed by geniuses who made their mark, with Maxwell and Einstein first and foremost, it really is a speed different from the others. It is a fundamental magnitude of the universe, invaluable to our understanding of space and time, subversive of established ideas and perceptions. Hidden behind the persistent attempt to measure its value accurately lay the eternal question as to the nature of light. A question which was only to be answered—and even then, only temporarily—by transcending its own logic.

2.6   Isaac Newton's Bright Ideas

Sir Isaac Newton, the first scientist to be honoured with a knighthood, is one of the greatest figures in the history of mankind. His life's work consisted of astounding accomplishments in physics, mathematics and astronomy. These do not merely include the laws governing gravity and the motion of bodies, expressed in elegant mathematics. They also contain something profounder: the view that the world is subject to general laws, and that the variety of phenomena is no more than the manifestation of those laws expressed uniformly, whether applied to the planets in the heavens or to a stone dropping to the ground. Newton's oeuvre stands as an overall view of the world, which does not seek to dig deeper into physics alone, but also into mathematics and theology.

Yet before he formulated the laws of mechanics and universal gravitation that are still to be found in school textbooks to this day (expressed so drily that pupils find them more off-putting than awe-inspiring), Newton worked persistently on light and its properties. This may have been because at that time, in the age of navigation and the telescope, improvements in optical instruments were much in need. But it is also possible that Newton grasped the vital importance of light in gaining an understanding of the world, just like Einstein later on and, more recently, Feynman.

By a curious trick of fate, Newton conceived of all his great ideas and the mathematical language with which to express them during an outbreak of the plague, which was decimating England in 1666. Having only recently completed his studies, he withdrew to the seclusion of his home. In 1669, at the age of just twenty-seven, he was appointed Lucasian Professor of Mathematics at the University of Cambridge. It's worth noting that in 1979—three hundred years later—the eminent theoretical physicist Stephen Hawking was appointed to the same chair, which he held until 2009. Hawking will go down in history not only for his scientific work, but also for the unimaginable courage he showed in battling against motor neurone disease.

Beyond their genius, there is perhaps something more Newton and Hawking have in common—their intuitive philosophical need for a comprehensive interpretation of phenomena. Newton expressed that need via the theory of universal gravitation and the laws of dynamics; Hawking did so via his attempt to unify gravitation and quantum mechanics, thus creating a "Theory of Everything". For the time being at least, it should be noted that the universe does not appear to obey this scientific or merely human desire.

As a university professor Newton experimented with lenses and prisms, teaching his novel ideas on the colours in light. Nevertheless, what paved the way to recognition and rare renown was the construction of the first "catoptric" or reflecting telescope. Using parabolic reflectors in place of lenses, Newton neutralised many of the aberrations typical of the Galilean telescope. This provoked the interest and unqualified praise of the powerful Royal Society in London, which elected Newton a member. It was to the Society that he addressed his first paper, in which he described his experiments and views on colours. Up until then, it had been believed that light was white in its initial form, and gained colour by modification, more or less in the way a white fabric is coloured by dyeing. Newton replaced this "modification" with the principle that light could be analysed into its individual constituents. In other words, he insisted that light was complex and heterogeneous rather than simple and homogeneous.

Though it now seems self-evident, the principle of analysis instantly gave rise to legitimate objections. Opposition was also voiced by Robert Hooke, a prominent experimenter and leading member of the Royal Society who was the author of *Micrographia*, an almanac of experiments and observations. Newton reacted vehemently, isolating himself for a long spell, while never ceasing to try and humiliate his detractor. Such behaviour was typical of him—Newton was in general hostile to criticism, rancorous and vindictive, and would often interlace his scientific disputes with demeaning conduct. One of his historic disagreements concerned the paternity of differential calculus, which is the mathematical basis of modern physics. A claim was made by the leading German mathematician and philosopher Gottfried Leibniz. As is now known, while Newton had invented calculus several years earlier, Leibniz really was the first person to publish a paper on the subject, quite independently. Such things are not unusual in the history of science. The unusual thing was that the Royal Society undertook to resolve the issue under the presidency of Newton himself, who appointed an investigative committee consisting of his friends. Besides, throughout the course of the dispute, articles in Newton's defence that bore the signatures of his young associates had actually been penned by the man himself.

Such human weaknesses may surprise those readers in the habit of mythologising great scientists or creators. Yet they are neither rare nor unaccountable. In the case of Newton, it would appear that an upbringing lacking in care played a decisive role in moulding his character. His father died three months after he was born; and while his mother remarried at once, she too departed this mortal coil when Newton was a mere three years old. His solitary, vulnerable personality led him to bouts of seclusion and resentment, even when he had gained fame and material prosperity. In fact, it is worth noting that Newton, the very epitome of scientific methodology, and to a great extent its creator, was deeply involved in alchemy and the occult, on which he wrote entire treatises in secret!

Newton's views on light gained considerable publicity via his lectures and talks at the Royal Society, but are only to be found in their entirety in his monumental *Opticks*. And while the focus of his papers and experiments concerns his theory of colours, reading between the lines of the *Opticks* one can see his fundamental understanding of light as being particulate in nature. "Are not light rays very small bodies emitted from shining surfaces?"[10] he wonders. Colours are attributed to the differing sizes of light particles and the difference in sense they produce in the eye.

In 1687 Newton published his core work, the *Principia*, or *Philosophiae Naturalis Principia Mathematica* in full, which stands as one of the high points in the history of science. It includes the law of refraction on the basis of the particulate nature of light. Though obviously inconsistent with reality, the conclusion drawn is that light travels faster in denser materials.

Publication of the *Principia* completed the overall schema of the natural world conceived by Newton in the plague years. As he writes: "I offer this work as the mathematical principles of philosophy, for the whole burden of philosophy seems to consist in this—from the phenomena of motions to investigate the forces of nature, and then from these forces to demonstrate the other phenomena [...] the motions of the planets, the comets, the moon, and the sea."[11]

On the basis of the laws of mechanics and universal gravitation, this grand schema unifies our image of the world, from the scale of the celestial bodies down to the smallest particles of matter. According to Newton, matter consisted of "solid, massy, hard, impenetrable, movable particles."[12] The crucial question was not what makes an apple fall to the ground, which is traditionally claimed to have led Newton to the theory of gravity. The essence of the question was whether that same thing held the moon in a fixed orbit around the Earth. "You had to be Newton," commented Paul Valéry, "to perceive that the moon falls, when everyone else could very well see that it doesn't fall."[13] On the other hand, J. Bronowski notes: "As a system of the world, of course, it (the *Principia*) was sensational from the moment it was published. It is a marvellous description of the world subsumed under a single set of laws. But much more, it is also a landmark in scientific method.

We think of the presentation of science as a series of propositions, one after another, as deriving from the mathematics of Euclid. And so it does. But it is not until Newton turned this into a physical system, by changing mathematics from a static to a dynamic account, that modern scientific method really begins to be rigorous."[14]

In that unified natural system, light was an entity that had to obey the same laws. Microscopic projectiles of light charted their course depending on the forces of attraction or repulsion they were subjected to; quite naturally, like all material bodies they travelled in a straight light if left to their own devices, according to the law of inertia. So for Newton, the particulate nature of light was a footnote to the overall picture of the world he had formed.

Particularly after the publication of his *Principia*, Newton's reputation and influence were perhaps beyond compare in the history of science. Admiration and praise for him began to extend beyond scientists to poets and artists. As Leonard Shlain has pointed out: "What began in the early Renaissance as a quickening in the understanding of nature culminated in 1687, when Newton published his all-encompassing *Principia Mathematica*, the Bible of the new scientific paradigm. Newton, continuing a theme begun by Descartes and Galileo, demoted God to the role of Grand Designer. [...] God ceased to be an active participant in the daily affairs of His subjects and became a passive observer of the creation He had set in motion."[15]

Be that as it may, for the last twenty years of his life Newton effectively abandoned scientific work to become Master of the English Mint. There he found a socially acceptable outlet for the persecution mania that haunted him, by sending several counterfeiters to the gallows. Newton departed this life, which is governed by the laws he himself discovered, in 1727. Born in 1642, he died in the fullness of years, full also of honours and recognition. Yet his paternalistic hold on the field of science lasted for at least a century beyond his death. As Alexander Pope's epitaph for him reads:

Nature and Nature's laws lay hid in Night
    God said, Let Newton be! and All was Light.

It is worth pointing out that Newton's halo of authority outshone any criticism regarding his views on the particulate nature of light, even though his theory suffered from substantial weaknesses and gaps. It was, for instance, incapable of offering a convincing explanation as to what happens when two people look at each other, in which case the bundles of particles would have to collide in motion. Over the years, the cracks in particulate theory became ever more apparent. Though no-one would have believed it, light appeared to have clear wave properties. It was in essence a wave, though one very different to the other waves known to nature.

2.7 Light Contributes to an Understanding of Itself

Many readers may well find it hard to see the connection between the behaviour of light and that of a wave. Yet the ripples familiar to us all when we throw a stone into a pond belong to a very wide range of phenomena. The world of physics, like the world we live in, is full of waves. The most traditional ones are mechanical waves, such as those seen in a spring, or those caused by earth tremors. Sound waves are commonplace. Strike the chord of a musical instrument, and sound waves spread in all directions. Readers may also be familiar with the important category of electromagnetic waves, including radio waves and TV waves, which can seriously damage our intellectual health.

For all their undoubted variety, wave phenomena share certain common characteristics. The main one is periodicity. In other words, a wave is a self-repeating displacement or disturbance. It is also easy to deduce that although this wave disturbance is propagated in space, the propagation medium does not follow it. Anyone fishing with a rod can see their float bobbing up and down as the waves come in, but the float—i.e. the water under it—does not follow the course of those waves.

As early as in Newton's time there were those who insisted that light was some kind of wave. The main proponent of the idea was Christiaan Huygens, another figure in late seventeenth century science. He made a major contribution to the development of dynamics and optics, and was even the inventor of the pendulum clock. In his *Treatise on Light*, published in 1690, Huygens concluded: "It is then in some other way that light spreads; and that which can lead us to comprehend it is the knowledge which we have of the spreading of sound in the air [...]; and consequently it spreads, as sound does, by spherical surfaces and waves: for I call them waves from their resemblance to those which are seen to be formed in water when a stone is thrown into it...."[16]

So a wave is a form or characteristic structure that moves. And while the wave view proposed by Huygens offered a convincing explanation for many of light's properties—above all refraction—his ideas remained marginalised in the shade of Newton. At any rate, they were not sufficiently backed up by mathematics. Not long afterwards, the pre-eminent mathematician Leonhard Euler was to formulate a comprehensive theory on the "oscillations" of luminous objects, by analogy with sound. But as often occurs in physics, established ideas were only overturned by experiment.

Firstly, like sound, we have seen that light is capable of being deflected by obstacles, or of spreading via narrow slits placed in its path. Such phenomena, known as diffraction effects, are only observed in waves—they cannot possibly be accounted for by particulate views. But what established the wave nature for good was interference, another property of waves confirmed via experiment. Waves of every kind, from ripples to sound waves, interfere. As the word indicates, this means that at some point in space, waves coming from different directions interlap, leading either to their amplifying or cancelling each other out. If that can be grasped for mechanical waves, in the case of light it leads to the following paradoxical consequence—sections of a uniformly lit screen may darken if the screen receives more light. In other words, light added to light may lead to darkness!

Proof that light interferes was provided in 1801, by one of the clever and delicate experiments performed by Thomas Young. In his youth he had shown a rare talent for advanced mathematics, the classics and natural philosophy, and by the age of 16 already knew twelve languages. It is thus hardly surprising that apart from physics and medicine, in which he received formal training, he showed great interest in archaeology. He even made a substantial contribution to deciphering the hieroglyphics carved on the famed Rosetta Stone.

By the time Young set to work on light, a new century had already dawned, and one that was to prove subversive of existing perceptions. As he proved via his experiments, under certain conditions the light emitted by two sources cancels itself out at particular points in space. This behaviour, which was due to so-called destructive wave interference, could obviously not be accounted for by the particulate theory. Light was indirectly but clearly hinting at its wave nature.

Though we may be unaware of it, interference effects have found a wide range of applications in the modern world, in the holograms that produce three-dimensional images of objects. Holograms are created by splitting a laser beam in two—one half illuminates the object in question, and is then reflected onto a photographic plate, while the other half is beamed straight at the plate. Complex interference effects on the plate or film create a hologram structure, which, when illuminated by another laser beam, creates the three-dimensional image. Hologram technology makes it possible to closely monitor changes in the texture or shape of materials caused by stress.

Another instance of how interference can be applied—though this time with natural light—is found in the credit cards familiar to us all. The small reflective surface they have in one corner makes use of interference to display logos and colours that show up in the right light. On the other hand, the consumerist madness typical of credit card holders is not so easily attributable to light interference!

We have already seen how the pioneer of wave theory was a giant of learning. What is more, Thomas Young was an intellectual with an appreciation for the finer things in life. A man of considerable prestige, he too was elected to the Royal Society at an early age, for correctly explaining how the eye changes to focus at different distances. Yet neither his prestige nor the truths revealed by his experiments (among other things, his wave theory accounted for the colours of thin films) would suffice to directly upset particulate views of light. His personal weaknesses may also have had a role to play in this; he was prone to derogatory outbursts, and his writing was often long-winded. Furthermore, as Charles Gillispie observes: "An important vein of resistance went deep into the very structure of classical physics. It is comparable to Newton's own inability to feel the full force of wave theory, and has to do with the influence on science of canons of style and language, which is to say of mathematical taste and technique."[17] So it took many more years, experiments and laborious mathematical operations for wave ideas to win the day.

An important role in their gaining supremacy was played by the French physicist Augustin Fresnel. He not only carried out extremely accurate experiments on diffraction and interference, but also devised a comprehensive mathematical framework for the wave theory of light. In parallel with Young's efforts, he even succeeded in employing wave theory to account for another phenomenon in optics—that of polarisation.

In today's world, the word polarisation brings to mind reckless blind-alley politics. But to physics, polarisation is a simple yet salient property of waves, which is manifested in various ways. For instance, light passing through special crystals is in some way "oriented" so that it comes out in polarised form. If a second similar crystal is then placed in its path in the right way, it can block the passage of light completely. Polarisation is responsible for the glare that drivers find irritating and sometimes dangerous when sunlight reflects off smooth surfaces; it is best dealt with by wearing so-called Polaroid sunglasses.

But polarisation was not simply yet another proof that light was a wave. More than that, it proved light was a transverse wave. In other words, light moved more like waves in the sea or a guitar string than sound waves, which are longitudinal, propagated in air and solid bodies by the successive compression and rarefaction of matter. On the other hand, the string of an instrument oscillates up and down or from left to right—transverse to the direction in which the sound is propagated. As a transverse wave, light is produced in the same manner—by oscillations perpendicular to the direction it is propagated in. That is also how polarization occurs, since polarizing materials only allow one oscillation direction through, so light passing through them emerges polarized.

Yet that was where other crucial questions destined to vex science arose. What on earth is it that oscillates to produce light waves? What is the material medium that enables them to propagate and travel? These were questions which, as has often occurred in the investigation of light, were to lead to fascinating, unanticipated worlds.

2.8   Aether Waves and Delusions

For the wave nature of light to stand up, it seemed self-evident that a material medium had to exist to convey light waves. This was particularly so when it turned out that sound—with which analogues had been sought—was propagated via matter, but not in a vacuum. By contrast, light had to cross the interstellar void to reach us from the distant stars. So the "void" in question could not exactly be empty; it had to be filled with some substance both material and invisible, so as to permit the propagation of light. This same material substance had to be contained in transparent bodies or the Earth's atmosphere, which are likewise permeated by light. It was a tall order, but not one that looked like daunting the physicists and mathematicians of the time.

This paradoxical substance filling space and matter, which was to be the propagation medium for light waves, had existed as an idea since ancient times: the aether, a luciform, divine substance linking everything together. The aether is a basic element in myths of profound cosmological and philosophical content. Greek philosophy later transposed it from a divine entity performing cosmological functions into a parameter of our world.

The idea of the aether was revived in the time of Descartes, this time on a scientific footing. Descartes rejected the notion of action at a distance without the mediation of a material medium. The aether was thus necessary for the propagation of gravitational and electrical forces. According to his view, then, the aether was thin matter permeating all bodies, whether astral or terrestrial, filling in the gaps in them. Heat and light were also transferred instantaneously via the aether.

Descartes' principal ideas on the aether, formulated in the first half of the seventeenth century, seriously influenced all subsequent views. To adapt to ever-increasing demands, the aether constantly gained properties that were as new as they were contradictory. It was originally regarded as an exceptionally thin gas, but then went on to acquire rigidity far exceeding that of steel! Everyone was certain of one thing: that the aether existed as a material entity. Its existence was essential for the propagation not only of light, but of all known forces; its impressive metamorphoses would even have surprised the Ancient Greeks.

In fact, the astonishing growth of mathematics was one new feature that complicated matters further still. Mainly from Euler onwards, there were in essence two languages for describing the natural world: that of the senses— the language of sight and hearing or, at any rate, of imagination—and that of the abstract, of mathematical calculus. Often imperceptibly, the profounder relationship between physical reality and the mathematical description of it is what has characterised the history of light. Indeed, in this particular case, where the aether was imposed as the propagation medium for light, the demand seems to have been unreasonable. At bottom, the object was to account for an invisible physical entity such as light with the aid of a similarly invisible and inconceivable entity, namely the aether.

After Galileo, however, physics evolved into an experimental science. Therein lies both its grandeur and its difficulty. There may have appeared to be a dire need for the aether to exist, and as early as the nineteenth century there were satisfactory mathematical schemas to describe its behaviour, but what remained was experimental confirmation of it. Concerned by the conflicting properties attributed to the aether, at a conference in 1889 the prominent physicist Heinrich Hertz was to stress the necessity of resolving "… the great problem of the nature and properties of the ether which fills space, of its structure, of its rest or motion, of its finite or infinite extent. More and more we feel that this is the all-important problem, and that its solution will not only reveal to us the nature of what used to be called imponderables, but also the nature of matter itself and of its most essential properties – weight and inertia…."[18]

So the systematic quest for the aether was undertaken by many laboratories in Europe, and by the burgeoning scientific community in America. Ingenious, highly accurate experiments were devised. All in vain. It seemed impossible to verify the existence of the aether.

2.9   The Importance of Being Wrong

In the late nineteenth century, two gifted American physicists named A. Michelson and E. Morley joined forces to investigate the evasive aether. Michelson was Polish by birth, but his parents had migrated to the United States, where he became a professor of physics at the University of Chicago. He had made a significant contribution to several branches of experimental physics, from atomic spectroscopy to astrophysics. He also spent considerable time on precisely measuring the speed of light, and in 1907 became the first American citizen to win the Nobel Prize for Physics. On the other hand, Morley had originally studied theology, but soon turned to the sciences. In the late 1860s he became professor of Chemistry and Natural History at Western Reserve College (later Case Western Reserve University), then in Hudson, Ohio. Ever possessed of an admirable passion for accurate measurement, he contributed both to precisely determining the percentage of oxygen in the air and to difficult atmospheric scale measurements.

The ingenious experiment devised by Michelson and Morley aimed to measure the speed at which the Earth moved though the aether. Yet the results came as a surprise, provoking years of heated debate; the reason behind the interest was not the experiment's success, but its unanticipated failure. For all the wisdom in its construction, and the experimenters' consummate skill, it was impossible to determine the motion of the Earth.

The idea behind the Michelson-Morley experiment, carried out in 1887, was simple. If a light beam was sent parallel to the Earth's direction of motion, the speed of light as measured would be increased by the Earth's speed. If, on the other hand, the light beam was moving against the Earth's motion through the aether, a terrestrial observer would see light propagated at a lower speed, since the Earth would then be heading to meet the light.

The paradox was that this logical reasoning did not appear to apply in the case of the Earth's motion through the aether. Michelson and Morley could measure no difference, whether light was travelling in the same direction as the Earth or against it. Sure enough, any differences would necessarily have been infinitesimal, since what would have been added to or subtracted from the colossal speed of light (300,000 km per second) was the speed of the Earth's rotation, which did not exceed 30 km per second. But the apparatus devised by the American physicists—the famed interferometer—was extremely sensitive. In it, light was beamed onto two mirrors placed at equal distances from the source point, but at right angles to each other. Reflected light returned from the mirrors, and the time taken for each of the two routes was compared. Even minute differences could be detected, thanks to the subtle interference effects that make extraordinarily accurate measurements possible.

Yet regardless of the direction in which the rays travelled, or the way the apparatus was rotated, light appeared to travel at the same invariable speed. There was no observable difference whatsoever. Other researchers repeated the experiment in many different versions, striving either to achieve greater accuracy or to pinpoint some error. In vain.

It was obvious that something was rotten in the state of Denmark. It was just that nobody knew what the state or Denmark really was. Could the guilty party be the speed of light, the interferometer measurements or the very motion of the Earth in the aether? Various attempts were made to reconcile the paradoxical results with the previously successful wave theory. For instance, it was argued that solid bodies moving in the aether contract in the dimension parallel to their motion; distinguished physicists went so far as to talk of some kind of "conspiracy of natural laws" that rendered motion relative to the aether undetectable.

An explanation for the conspiracy was given by Einstein and the special theory of relativity, though it was one both unexpected and radical: what was rotten lay in our perception of time and space. It follows that Einstein was not referring to the Michelson-Morley experiment. His route was a different one; he was interested in the building's foundations, not the cracks in the walls. So although one of the most important experiments in the history of physics was easily accounted for by relativity theory, the former had little if any hand in the conception of the latter! As for the aether, it didn't even exist. Its presence was not only unnecessary, but even appeared incompatible with the new ideas brought by relativity.

Today, over a century later, these findings seem more than obvious. So much so, indeed, that it is difficult to comprehend the pernicious insistence on the notion of the aether. It would be easy to argue that the Michelson-Morley experiment was not even worth doing, to say nothing of the meticulous provisions made. But as mathematician and cosmologist Hermann Bondi rightly stresses: "Such hindsight is not useful in science. What makes the Michelson-Morley experiment so celebrated is that it first proved something that has so deeply entered our thinking that it has become obvious."[19] A valuable dictum, applicable well beyond the bounds of physical science. Many "obvious", stereotypical notions dominate our thought, and sometimes regulate our lives. It might be worthwhile questioning their value, and the deeper-seated causes they may conceal.

That said, it is worth noting that the idea of the aether resurfaces in modern cosmology, though of course with a different content. It is linked to cosmic microwave background radiation, one of the greatest discoveries in the history of science. In essence, this cosmic radiation is weak light. It fills the entire expanse of the universe, reaching Earth from the outermost boundaries of space and time. It is as if light coveted the glory once accorded to the aether, and is striving to persuade us that it, on the other hand, is no illusion.

2.10   The Fields of an Autodidact

In the eighteenth century, light and its colours, improvements in lenses and telescopes, and the properties of the aether—in short, optical phenomena—were widely studied by both professional and amateur scientists. But at the same time, the radically different category of electric and magnetic phenomena had begun to gain in popularity. These covered a broad spectrum, in a field enriched with major discoveries by maverick personalities. Nevertheless, it was the outstanding figure of Michael Faraday who, despite being self-taught, slowly but surely enriched our experimental knowledge of electricity and magnetism. Above all, he was to link those two apparently different entities.

Even to this day, it seems daring to contemplate and difficult to explain the fact that magnetic attraction of the kind exerted by the Earth's field on a compass is related to electricity and its numerous manifestations. But the experiments carried out by Faraday and others in the first half of the nineteenth century leave us with no doubt. As early as 1820, the Danish physicist Ørsted observed during one of his classes at the University of Copenhagen that an electric current passing through a platinum wire would deflect a compass needle. This was the birth of electromagnetism, before what were doubtless the glazed eyes of his students.

The next major contribution lay in the experiments and theoretical papers of the talented French mathematician, chemist and physicist André Ampère. Here we may simply note that in following the educational trends of the time, which were probably a great deal better than those nowadays, Ampère owed his basic education solely to his father's well-stocked library.

Nevertheless, the truly groundbreaking discovery in electromagnetism was made later, by Faraday. To his surprise, he determined that electric current is generated whenever a conductor moves near a magnet, and vice-versa. The same thing also occurs when the electricity flowing through a coil alternates; electric current then appears "by induction" in a second coil placed nearby. According to Faraday, "waves of electricity" produced in the first coil travel in space, causing a similar disturbance in the second. This was the manifestation of an important phenomenon that became known as electromagnetic induction. Though such a fancy name is bound to be off-putting, the discovery of induction not only found important applications, but was destined to change our ideas about the very nature of light.

Over the ensuing decades, Faraday passionately pursued his masterful experimental study of electromagnetic effects. He was also to introduce the theoretical concept of the "field", effectively inaugurating a new era in physics, in which the mysterious forces acting between bodies at a distance were replaced by the field. This was distributed uniformly in space, and represented by force lines. Both Faraday's conception of force lines and the notion of the field were obviously empirical and, to some extent, intuitive. But given a mathematical foundation, electrical and magnetic fields were, like gravitational field, to emerge as fundamental entities of the physical world.

Interestingly, neither Faraday's education nor his path in life presaged the major contribution he made to numerous scientific "fields", in the more usual sense of the word. Born into a poor English family in 1791, he was forced to abandon his schooling while still young to become apprentice to a bookbinder. In the present era of often profligate education, the way he himself describes gaining his early knowledge is instructive: "Whilst an apprentice I loved to read scientific books which were under my hands, and, amongst them, delighted in Marcet's *Conversations in Chemistry*, and the electrical treatises in the *Encyclopaedia Britannica*. I made such simple experiments in chemistry as could be defrayed in their expense by a few pence per week, and also constructed an electrical machine, first with a glass phial, and afterwards with a real cylinder, as well as other electrical apparatus of a corresponding kind."[20]

At the age of 22 Faraday was taken on at the Royal Institute, as assistant to the celebrated contemporary chemist Sir Humphry Davy. He did not earn this humble position so much out of his passion for science, as on account of his unmatched skill as a calligrapher and book binder: having managed to attend Davy's lectures, he sent the chemist a finely bound volume of his notes. In any case, Faraday was to remain at the Royal Institute for the rest of his life, first as a factotum, later as a fellow and eventually as Davy's successor, when he passed away.

Faraday's fame grew by leaps and bounds, and he won scientific accolades all over the globe. He was also offered a knighthood, and the presidencies of the Royal Institute and the Royal Society, all of which he steadfastly declined. "I must remain plain Michael Faraday to the last," as he declared.

So it was that around the middle of the nineteenth century, experiments, observations and theoretical explanations had amassed a great wealth of knowledge—and without doubt no few contradictions—on electrical and magnetic effects. The relationship between them was plain; as so often occurs in science, the atmosphere seemed ripe for the daring quantum leap. It was as if the play, the cast and the scenery had all been tenaciously prepared, but the only things missing were the director and his mind and talent, to breathe life and meaning into the script.

2.11   A Bolt from the Blue

The conviction that light consisted of waves due to the oscillations of the aether, an almost phantom yet material entity, had created a comprehensive picture free of substantial gaps. But that picture was to be shattered by a bolt from the blue. Via a different route, experimental indications convincingly argued that electricity and magnetism were interdependent phenomena awaiting theoretical unification. The mind destined to unite the piecemeal knowledge about the two already existed. It had appeared in 1831, a mere two months after the discovery of electromagnetic induction. And as circumstances demanded, it was the mind of a mathematical genius: that of James Clerk Maxwell.

By the age of sixteen Maxwell was already a student at the University of Edinburgh, where he studied physics, mathematics and philosophy. In 1856 he became a professor, on his way to a brilliant career. He formulated the kinetic theory of gases, accounted for the rings of Saturn and studied composite colours. First and foremost, however, he was the father of electromagnetic theory. The publication in 1864 of his paper on "A dynamical theory of the electromagnetic field" marked a milestone in the history of science. In his own words: "The theory I propose may therefore be called a theory of the Electromagnetic Field, because it has to do with the space in the neighbourhood of the electric or magnetic bodies, and it may be called a Dynamical Theory, because it assumes that in that space there is matter in motion, by which the observed electromagnetic phenomena are produced."[21]

In just four neat equations, Maxwell summarised the knowledge acquired about electricity and magnetism, gave a theoretical explanation for experiments and opened up the way for future applications. As the eminent Greek physicist E. Economou stresses: "The thousands of matters related to the production and transmission of electrical energy; the thousands of phenomena linked to electricity consumption in homes, factories and offices; the numerous issues of cable and wireless communication, from the telephone to the television; light and associated optical effects; X-rays; the forces determining atomic, chemical and biological phenomena, as well as those determining the properties of solid, liquid and gaseous bodies; thunder and lightning: all of these phenomena and many more can be accounted for, predicted and monitored on the basis of just four fundamental equations – Maxwell's equations."[22]

In effect, the equations unify electricity with magnetism. They show that when altered, every magnetic field results in the production of an electric field and vice-versa. Indeed, the rate of alteration determines the distribution of the electric or magnetic field in space; there is a quantitative relationship between the two, and not merely a qualitative one.

Thus, if we know how charged conductors, magnetised bodies or electric currents are distributed, we can use Maxwell's equations to precisely calculate the electromagnetic field in the surrounding space. It follows that bodies are not immersed in material aether, but in an electromagnetic field. The lines of force from this field extend out into space, surrounding and penetrating the bodies they encounter. Via this synthesis, a unified interpretation is effortlessly acquired for magnetic and electrical phenomena such as induction, electrical charge attraction and compass needle deflection. As Richard Feynman said: "From a long view of the history of mankind—seen from, say, ten thousand years from now—there can be no doubt that the most significant event of the nineteenth century will be judged as Maxwell's discovery of the laws of electrodynamics. The American Civil War will pale into provincial insignificance in comparison with this important scientific event of the same decade."[23]

Of course, the question then arises as to the speed at which electromagnetic waves travel in space. Maxwell did his calculations on the basis of pre-existing electrical and magnetic measurements. The conclusion was sensational: electromagnetic waves were propagated at the same speed as light! In other words, light too was an electromagnetic wave. As Maxwell himself stresses: "The agreement of the results seems to shew that light and magnetism are affections of the same substance, and that light is an electromagnetic disturbance propagated through the field according to electromagnetic laws."[24]

That simple sentence encapsulates an extremely profound change in our image of light. From then on and forever more, magnetism, electricity and light were inextricably bound up. Sunlight, for instance, or a laser beam are travelling forms of electrical and magnetic waves. An unexpected answer had at last been found to the tantalising question of what light waves consisted of—they were electrical and magnetic fields. When writing his famed equations, it was as if Maxwell were lending a mathematical interpretation to the biblical command "Let there be light!"

If you are inquisitive, you may remember that at that time the aether was seen as vital to the propagation of light. No need to worry—Maxwell believed the same himself. Indeed, as he enthused: "The vast interplanetary and interstellar regions will no longer be regarded as waste places in the universe, which the Creator has not seen fit to fill with the symbols of the manifold order of His kingdom. We shall find them to be already full of this wonderful medium; so full, that no human power can remove it from the smallest portion of space, or produce the slightest flaw in its infinite continuity. It extends unbroken from star to star..."[25] Note that Maxwell believed in the existence of the aether without using it in his calculations or having any need of it in the equations he formulated.

Maxwell departed this life in 1879. A few years previously he had collected his papers in the monumental *Treatise on Electricity and Magnetism.* Both physicists contemporary with him and succeeding generations regarded the book as an important yet inaccessible historical document. Maxwell's theory spread across Europe via its ever-increasing applications and by those who continued his work. Bearing in mind the preceding "aethereal" impasse and the semantics involved, the electromagnetic nature of light really did come as a bolt from the blue.

2.12   The Dream Begins and Ends

It is now worth taking a deep breath, and looking at how our image of the world changed once electromagnetic theory took hold. It was a radical change, leading from a materialist world governed by Newton's wondrous laws, to a fleeting one dominated by the presence of electromagnetic waves. After centuries of searching, light proved to be but a small part of the wonders of electromagnetic radiation, which was ubiquitous, being of various forms and origins.

Technological applications of electromagnetism were in the offing, destined to become the main feature not only of everyday life, but of modern civilisation. Yet the philosophical and cultural repercussions borne by the triumph of electromagnetic theory were no less significant. As Einstein observed: "The greatest change in the axiomatic basis of physics—in other words, of our conception of the structure of reality—since Newton laid the foundation of theoretical physics was brought about by Faraday's and Maxwell's work on electromagnetic phenomena."[26]

There can be no doubt that the end of the nineteenth century found physics at a triumphant watershed. The world had been interpreted on the basis of a small number of equations in classical mechanics, and a similarly small number of equations in electromagnetism. In Newton's laws, the motion of the moon, the trajectory of a stone and the inertia and acceleration of a body found a profound, austere descriptive framework. On the other hand, Maxwell, Faraday and Hertz—the fathers of electromagnetism—revealed that a vast wealth of phenomena, from the colours of the prism to the magnetic field surrounding a conductor, were subject to the same unifying theory. Two forces of known behaviour, gravity and electromagnetic force, were all that it took to comprehend the world. Indeed, since the notion of field was turning out to be more profound and substantial than that of force, the classical world of physics ended up being characterised by just two fundamental force fields: gravitational field and electromagnetic field. An endless variety of empirical and experimental observation had been condensed into an amazingly small number of fundamental equations and concepts.

Though some problems did of course remain unresolved, they resembled mere details, trivial imperfections on the face of an imposing edifice. Besides, the general conviction was that those problems would rapidly be overcome, albeit with minor adjustments here and there. The edifice itself was in no danger. In characteristically optimistic tones, A. Michelson was to write in 1899 that: "The more important fundamental laws and facts of physical science have all been discovered and are now so firmly established that the possibility of their ever being supplanted in consequence of new discoveries is remote."[27] All the same, it is worth stressing that the way towards that unanticipated upset was paved by the experiment that Michelson himself had designed to measure the speed of the Earth through the aether.

And though the electron had already been discovered, the then emerging world of atoms was no cause of concern for physicists. It seemed perfectly reasonable that this world, as part of a whole, should be governed by the same laws. All that remained was experimental confirmation of that self-evident truth. Many eminent physicists of the time thus advised the young not to go into physics, since nothing worthwhile remained to be investigated. In around 1900, Lord Kelvin is said to have argued that all the fundamental problems in physics had been solved, and that it only remained for general laws to be applied. There were of course a number of minor matters still not fully understood, such as black body radiation and the spectral lines of gases. In an oblique reference to Kelvin, Michelson noted how: "An eminent physicist has remarked that the future truths of Physical Science are to be looked for in the sixth place of decimals."[28]

These delusions regarding a full understanding of the world were of course entirely compatible with the complacency typical of contemporary Victorian society. The middle and upper classes believed they had created a more or less perfect society that would last forever. Some minor social problems still remaining would soon be resolved. Indeed, impressed by the triumph of the theoretical principles of physics, intellectuals of the time went so far as to believe that everything in the world could be accounted for with their own contribution. And since the laws of classical physics were absolutely deterministic, if one could record a particular moment in the motion of all the atoms in the universe, then one would know both the past and the future. Thus for the great mathematician Laplace, the universe was an unerring machine, a cosmic clock set in motion at some point. Ever since, history had been predetermined down to the very last detail: social conflicts, the fall of an empire and the course of a love affair were simply due to the unstoppable action of physical laws at the ultimate atomic level. Fortunately, recording every atom and its motion appeared practically impossible. Anything different would have meant omniscience—a prospect that was appealing to Laplace, but nightmarish as a possibility. Besides, the fleeting presence of electromagnetic fields had dealt the simplistic, mechanistic view of nature a severe blow.

At any rate, the triumph of classical physics even appears to have influenced such major thinkers as Marx and Freud. A curious social determinism held sway. This was taken to extremes when, in defending a criminal, a celebrated lawyer of the time invoked the inescapability of his actions. As he claimed, these were the simple consequence of a primordial mechanism set in motion at some immemorial time, without the criminal's volition or culpability!

Luckily enough, the evolution of physics argued otherwise. Thus, at the dawn of the twentieth century, the microcosm of atoms, the nucleus and electrons passed rapidly from the theoretical hypothesizing stage to that of experimental research. The discovery of X-rays and radioactivity, the development of spectroscopy and the behaviour of the electron led to unanticipated cracks appearing in the edifice of classical physics. Nor could any convincing explanation be offered for the distribution of radiation emitted by a heated body, while the inability to measure the speed of the Earth in the luminiferous aether by experiment was a continuing cause for concern.

These "minor matters", as Lord Kelvin's dismissive turn of phrase would have it—in other words, certain experimental results that could not be satisfactorily accounted for in terms of classical ideas—were unexpectedly to lead to two major revolutions: the special theory of relativity and quantum mechanics. From the very outset, we should point out that like all revolutions, they did not do away with the established order in its entirety. They changed the way we see the world, added to our computational tools and created the framework for the explosion of technology so typical of our times. Nevertheless, in the everyday world Newtonian dynamics remained valid, and electromagnetic theory did not cease to describe the way a wave is produced and propagated. But what we now knew were the limits, inherent weaknesses and impasses of classical ideas.

Interestingly, the phenomena destined to shake our confidence in the classical image of the world all concerned light and its properties. Trapped in the bounds of the classical world, it was as if light were bent on issuing a warning and a protest. In other words, it was in a hurry to hint at its central significance in comprehending the universe.

3

Light's Revelation of Time and Space

3.1 Einstein's Magic Carpet

As we have seen, the twentieth century began what would prove to be its iconoclastic life certain that the physical interpretation of the world was more or less over and done with. Newtonian dynamics governed the motion of solid bodies, from the planets to the atomic constituents of matter. On the other hand, electrical and magnetic effects obeyed Maxwell's neat equations, with one of the consequences being that electromagnetic waves were propagated at the speed of light. These grand concepts of Newtonian mechanics and electromagnetism, which found numerous applications, interpreted individual phenomena and employed specialist mathematics, were what made up the classical edifice of physics. As is obvious, within its framework the world was separated into two diametrically opposed and seemingly incompatible entities: matter particles and electromagnetic waves. Through this logical separation into particles and waves, classical physics had acquired remarkable unity and economy.

Be that as it may, a bridge linking the two entities already existed in the shape of the aether—a somewhat exotic material medium, yet one which appeared necessary for the propagation of electromagnetic waves. Conventional wisdom said it was only a matter of time before it was confirmed by experiment.

Consequently, the sky fashioned by physics in the late nineteenth century was clear and tranquil. A few small clouds in the form of paradoxical experimental results gave no forewarning of the storm about to break. One of

© The Author(s), under exclusive license to Springer Nature
Switzerland AG 2024
G. Grammatikakis, *The Autobiography of Light*,
https://doi.org/10.1007/978-3-031-56917-3_3

those clouds concerned the carefully conducted Michelson-Morley experiment, aimed at determining the motion of the Earth in the aether. Curiously, no such motion could be detected.

At the end of the nineteenth century, however, a somewhat malicious Albert Einstein lay in ambush against the classical notions in physics, with an almost impalpable smile on his face. His own road was an entirely different one; rather than minor outstanding issues, he was interested in a profounder understanding of how the world functioned. With the special theory of relativity, he became the first person to shake the composure and certainties of classical physics.

This radical theory was propounded in the early twentieth century, while Einstein was still a lowly clerk in a patent office. Characteristically, however, what led the genius physicist to his theory were questions that once again had to do with light. Indeed, in the first instance they sound simplistic. But just as in life, apparently innocent questions are occasionally of great importance for science. They are capable of getting to the root of a problem, and enable us to pinpoint the weaknesses in established ideas. In other words, they lead the way to a necessary watershed in scientific thinking.

Indeed, it is often the case that once propounded, an entire thought experiment can lead to fruitful reflection. As its name indicates, a thought experiment is practically impossible to carry out in reality, yet neither its conception nor its evolution violates any physical law. Any number of fundamental questions have been explored by thought experiments, provoking heated debate and controversy. That was Albert Einstein's favourite game, a game that was to prove far from innocent. It reflected his deeper nature, one in search of the essential truth often hidden behind deceptive phenomena.

So when Einstein began thinking about light he did not rely on any experimental results, nor did he adhere to prevailing theoretical notions. As he was to do throughout his scientific career, in a curious way he rose above and beyond them. He sought his own thread, which was to lead both him and science out into a true clearing in the forest. Whether preoccupied by the speed of light or the nature of gravity, Einstein always appeared interested in the Entirety. "It seems that the human mind has first to construct forms independently before we can find them in things,"[1] as he said.

From the time when, as a sixteen-year-old schoolboy, Einstein had begun to think about light, one question buzzed constantly round and round in his mind: what would happen if somebody travelled with light, say by getting on a magic carpet that could fly at the speed of light? From that carpet, what would light itself look like, or the world, or a nearby clock disinterestedly measuring time?

The truth is that these questions are deceptive in their simplicity. If we really do rely on everyday life experience or even on Newton's precisely formulated laws, the answer to them appears to be paradoxical and nonsensical. And that is because if someone were to travel on a magic carpet moving at the speed of light, then light would appear to stop! At least, that can be surmised from similar experiences. On a turbulence-free flight, both the passenger next to us and the aircraft itself appear stationary, since we participate in their motion. But something entirely different must occur with light, since stationary light is obviously inconceivable.

Things look even more paradoxical if we extend the thought experiment into time. A railway station clock obviously shows the same time, regardless of whether it is consulted by the station manager from his platform or a by a passenger on a rapidly passing train. But an unconventional traveller who left for better worlds by boarding an imaginary light rocket would see the clock constantly showing the same time—precisely when the rocket was launched. And that is because the light waves reaching the traveller's eyes from the clock would constantly follow her, whereas the new waves continually setting out from the clock and showing time passing would never reach her.

There is thus an unexpected link between light, its speed and the notion of time. So however innocent and obvious they seem, Einstein's mental games led to the overthrow of ideas that had for centuries been regarded as unshakeable. As Alan Lightman characteristically puts it in his book *Einstein's Dreams*: "In the long, narrow office on Speichergasse, the room full of practical ideas, the young patent clerk still sprawls in his chair, head down on the desk. For the past several months, since the middle of April, he has dreamed many dreams about time. But the dreaming is finished. Out of many possible natures of time, imagined in as many nights, one seems compelling. Not that the others are impossible. The others might exist in other worlds."[2]

As the special theory of relativity was to show, in our world time does not pass independently of an observer's motion, nor does it remain indifferent to space in all three dimensions. Instead, it is space's fourth dimension! Einstein's magic carpet proved more magical than those in folk tales.

3.2   A Lone Traveller Through the Universe

The paradox of travelling along with light was obviously not the only thing that led Einstein to his astonishing conception of the theory of relativity. While he matured as a scientist, both the existence of the aether and the nature of time and space were constantly revolving in his mind. He was born in the German city of Ulm in 1879, into a family of no particular scientific or intellectual distinction. After some repressive years at an authoritarian school in Munich, he left, unable to tolerate the prevailing climate of discipline and control, to which he never reconciled himself. He then struggled to gain entry to Zurich Polytechnic. His student years passed by indifferently, without revealing any of his talents, and he apparently managed to graduate by borrowing notes from a conscientious student. As he himself wrote: "In this field [physics], however, I soon learned to scent out that which was able to lead to fundamentals and to turn aside from everything else, from the multitude of things which clutter up the mind and divert it from the essential. The hitch in this was, of course, that one had to cram all this stuff into one's mind for the examinations."[3] Anyone who has had a taste of university education would obviously vouch for that bitter comment.

Be that as it may, behind Einstein's dreamy and often absent-minded gaze burned the unquenchable need to better understand the world. Much of his time was spent reading original, complex texts in physics and mathematics. Less well known is the fact that he, the patriarch of theoretical ideas, liked devoting time to laboratories and experiments. In fact, he even joined his friend Leo Szilard (also a leading physicist) in designing an innovative fridge with no moving parts, which might have gone into production had it not been for the outbreak of World War II.

When Einstein graduated without distinction from Zurich Polytechnic in 1900, he was forced to work at the Patent Office in Bern to make his living—and even then it took a family friend to pull strings. It obviously sounds like a crying shame that one of the greatest geniuses ever born should have worked on patent applications and licences, such as one for an improved type of double-barrelled shotgun, or another for a new way of controlling alternating current, which he judged to be "incorrect and imprecise". But the truth of the matter is that this humble position left Einstein with plenty of time for his own interests—in any case, he had never had much time for hierarchies and formalities.

It was in 1905, in this scarcely academic environment, that four of the greatest papers in the history of physics came into the world. Considering that they both literally and metaphorically saw the light far from the bosom of academia and research, 1905 was quite rightly termed an *annus mirabilis* for science.

One of the papers Einstein published in 1905 contained the answer to the paradox of a passenger travelling at the speed of light. Destined to become known as the "Special Theory of Relativity", this answer overturned all classical, familiar perceptions of time and space. "Page for page," observes Arthur

I. Miller, "Einstein's relativity paper is unparalleled in the history of science in its depth, breadth and sheer intellectual virtuosity. Einstein developed one of the most far-reaching theories in the history of physics in a literary and scientific style that was parsimonious yet not lacking in essentials [...]."[4]

All the same, the relativity paper lay outside the current of the times, and initially went almost unobserved. In fact, when Einstein submitted it to the University of Bern two years later to apply for work as an assistant, it was described as "incomprehensible". But with support from a number of contemporary scientists—not least Max Planck—who judged his articles to be ingenious, Einstein's fame was not long in spreading. In 1909 he was offered an honorary position as associate professor in Theoretical Physics at the University of Zurich. This marked his triumphant entry into the world of academia, though he was never to identify with its arrogance and frequent authoritarianism. After some intermediate steps in university hierarchy, in 1914 Einstein accepted the research professorship especially established for him at the Humboldt University of Berlin; apart from anything else, this gave him leeway to teach whenever he wished. This was at a time when the university had brought together some of the finest minds of the times. It is even said that on his way to Berlin he stopped off at his family home and old school. Not deeming his former pupil capable of anything, the principal thought that he had come to beg!

A few years later, Einstein was to publish his work on the general theory of relativity—a radical interpretation of gravity that became the bible on the structure and evolution of the universe. This won him global fame, shrouding his personality in an aura of myth and hyperbole. His papers on relativity were still a cause for circumspection nonetheless, so in 1921 he was awarded the Nobel Prize in Physics "for his discovery of the law of the photoelectric effect". Via this discovery Einstein paved the way for the quantum physics revolution, only to become a vehement opponent of it later on.

Germany, however, was no longer the same. Hitler's ascent to power in 1933 led Einstein to renounce his German citizenship and settle in the United States. There he was to spend the rest of his life, at the famed University of Princeton. Yet his contribution to science had more or less come to an end. The quest for a Unified Theory encompassing all of the forces of nature constantly preoccupied him, but did not prosper. Alongside this he poured his energies into pacifism, without avoiding what was often a naïve engagement in matters political. In 1952 he declined an offer to become President of Israel, confirming his abhorrence of authority in any form. For all his great fame, he was to remain a humble and essentially solitary figure. As he himself wrote: "Perhaps, someday, solitude will come to be properly recognized and appreciated as the teacher of personality."[5]

Einstein's family life was, in any case, never cloudless. From 1903 he was married to Mileva Marić, a fellow student of Serbian Greek Orthodox descent. Though they spent seventeen whole years together, Einstein himself confessed that he never really got to know her. Nor was his own stance towards her beyond reproach. While still married to Mileva and living in Zurich, Einstein began a romantic attachment at a distance with his cousin Elsa. Their relationship became official when he remarried in 1919, after a far from amicable divorce from Mileva. From then on Einstein was to know many sorrows, culminating in Elsa's protracted illness and eventual death in 1936.

The difficulties in Einstein's personal life, above all from his first marriage, were to accompany him into old age. Dedicated to the riddles of science, his distant personality no doubt contributed to this. Ever since the publication of his personal archive, which contains hundreds of love letters among other things, a kind of literary genre possibly inspired by current scandalmongering has focused on Einstein's personal life and his failings. Yet it is precisely the most intimate, painful moments in his life that render his scientific achievements and constant attentiveness to the world so wondrous. A superhuman genius would have been repulsive. But he was a human, whose moments of weakness shed ever more light on the true scale of his genius.

Einstein was to end his days peacefully at Princeton in 1955. Shortly beforehand, he is said to have noted: "I have no special talent. I am only passionately curious." Nevertheless, the major scientific accomplishments of our time increasingly serve to underline his unique contribution to understanding the world. Einstein was a lone traveller through the universe. Perhaps that is why he sensed its rhythms more than anyone else.

3.3   Simple Axioms with Major Consequences

If unsuspecting readers could study the special theory of relativity, we can be sure they would be astonished at its simplicity and apparent innocence. No need to worry: the scientific community of the time felt much the same, used as it was to long-winded, complex calculations arising from Newton's laws or Maxwell's electromagnetic theory. By contrast, the entire magnificent edifice of relativity rests on two basic postulates. These take the form of axioms, in the profound Ancient Greek sense of the term. But before expounding them in detail, it is helpful to draw on some thoughts from experience—or at least from what we learned at school.

First of all, it is easy to deduce that physical laws always appear to remain the same in systems moving uniformly relative to each other. Take, for instance, a ship travelling in calm waters or a train slowly crossing a plain. Regardless of whether the description of a phenomenon is based on the ship, the train or some other "stationary" system, the laws of nature remain unaffected. So a ball bounces in the same way—i.e. following the same laws—on the ship, the train or the ground. A clock pendulum swings in the same harmonic motion at the same rate whether it is in a cabin on board the ship or on the quay. Even in a jet plane travelling at 600 km per hour, coffee spills from a cup just as it does in our living room at home.

That all seems more or less normal. In fact, it had first been formulated by Galileo in the first principle of relativity, which was incorporated into the grandiose edifice of Newtonian mechanics. Yet the fine point that Einstein focused his thought on was that the immutability of nature's laws applies right the way across physics. In other words, it is not limited to mechanics, but also includes the phenomena of electromagnetism and optics. And sure enough: providing there is no turbulence, a laptop or a digital voice recorder work the same way on board an aeroplane as they do on the ground. Yet in order from them to work, both of those attractive gadgets rely on laws belonging to different branches of physics: not only on electromagnetism and mechanics, but also on acoustics. If those laws changed in a moving system, the gadgets would have trouble working.

The first axiom of relativity encompasses this brilliant thought. According to its strict precepts, the laws of nature apply with the same content and mathematical form in all systems moving uniformly, i.e. at a constant speed relative to each other.

But as we have said, the new step that once again underscored the unity of physics was that the laws of electromagnetism also obeyed the principle of immutability. Thus light, being propagated like an electromagnetic wave, had to obey the same laws. And sure enough: it is just as easy to read and chat to your neighbour on an aeroplane as it is on the ground. Yet reading is connected to light and how it is propagated, whereas chatting is a mechanical phenomenon, relying on the propagation of sound through air.

The first obvious conclusion to arise here from the special theory of relativity was that the renowned aether had no reason to exist; further still, it could not possibly exist. The aether was contrived to account for the propagation of light. In other words, it played the role performed by air for sound. However illogical it now sounds, the aether's existence would result in optical effects on an aeroplane differing in form from those on the ground. Given that the aether was believed to be motionless relative to the ground, an aether wind would blow inside the aeroplane, influencing the propagation of light. But if that was the case, then we could use appropriate optical experiments to distinguish between a state of uniform motion and one of rest. And that runs contrary to the first axiom of relativity.

Along with the aether, the notion of an absolute frame of reference also died out. All motion is relative, though only with regard to arbitrary frames of reference. There is no motionless position or frame fixed in the universe. A rocket cannot measure its speed relative to empty space, but only relative to other rockets or the Earth. We also know from experience that when we see a train on the track next to ours, it is difficult to work out which one is moving: is it ours, the other one or perhaps both? Whatever the case may be, the second train is travelling at one speed relative to us, and another relative to someone watching at the station. Even winds can decrease or increase in speed depending on whether we are travelling with or against them.

Speeds, then, are relative. Yet though every speed has meaning in relation to a given frame of reference (which is always arbitrary), one important exception has earth-shattering consequences: the speed of light. It was precisely the second axiom of relativity that neatly expressed that inviolable law. Consequently, the speed of light always maintains the same value, regardless of how the source emitting it or an observer are moving. In other words, the speed of light is constant and immutable under all circumstances. It is a universal constant.

By any measure of common sense and experience, this axiom sounds preposterous. It is like measuring the speed of a bullet and always finding it the same, regardless of whether it is fired from a stationary gun, or the same gun fitted to a supersonic aircraft. So how can such an illogical law possibly apply to the speed of light? And yet it does! Whether stationary on the ground or installed on a rocket, a device measuring the speed of light from a distant star would always give the same value—300,000 km per second. Even if the star is constantly moving away as the universe expands—as has been shown to occur in the case of distant galaxies—its light reaches Earth at the same consistently fixed speed.

Yet however paradoxical all of this may seem, the truth is that classical physics also led to peculiar impasses as far as thought journeys with light were concerned. For instance, if a passenger on a rocket travelling at the speed of light were to hold up a mirror and look into it, his image would disappear! And that would occur because the light leaving his face would never be able to catch up with the mirror, since it would be moving at precisely the same speed. "Damn! There goes my image again," exclaims a rocket passenger in a comic book on relativity. "I keep telling them not to go 186,000 miles-per-second when I'm shaving."[6]

However, this illogical situation does not apply in the brave new world constructed by relativity. And that's because if the image of a speed of light traveller did disappear from the mirror, there would be a simple way of determining he was moving, which is something the first principle of relativity forbids. It follows that the image would stay in place. Indeed, while the passenger would know that light was leaving his face at the highest possible speed in nature, observers on the ground would measure the same speed—but if ignorant of relativity, they would continue to believe that the passenger couldn't have a proper shave.

So faced by Einstein's youthful question on how a light beam would look to someone travelling with it, classical physics replied that it would appear motionless. This answer quite rightly sounded irrational. On the other hand, the axioms of relativity led to an equally irrational answer: an observer travelling at the speed of light would yet again see light pulling away at the speed of light!

These paradoxes were only solved by an upset as radical as it was unexpected. And that upset concerned the notion of time.

3.4   Brilliant Answers to Old Questions

The truth is that at first glance, the axioms of relativity appear to be innocent and devoid of any special significance. As we have already said, the first stresses the immutability of physical laws in all frames of reference moving uniformly; the second regards the speed of light as constant and non-negotiable, irrespective of whether its source or observer are moving. Yet for all their simplicity, careful investigation of the axioms unexpectedly sheds light on how our world works. Furthermore, it lends different content to the laws of that world.

To start with, we must be aware that the speed of light is a speed like none other. It is, firstly, the fastest speed that does or indeed could possibly exist in nature. It marks the outer limit that nothing—neither material body nor radiation—may exceed or even reach. And however paradoxical it may seem, light maintains the same strictly determined value. Who or what determined it is another question. As we have seen, even if measured by a fast spaceship racing in the direction of a light beam, the speed of light is again found to be constant, at 300,000 km per second!

Light, then, is a solitary, peculiar traveller, which fate has destined to move in space at an ever-constant speed, conveying information and revealing the world. According to this daring rationale, the failure of the Michelson-Morley experiment now seemed self-evident. As we have already mentioned, the experiment aimed to ascertain the motion of the Earth through the stationary aether by using optical methods. However, such motion would have been impossible to ascertain on the basis of the axioms of relativity, since the speed of a light beam would remain unaffected by the motion of the Earth or the direction of the beam. Einstein's first axiom tossed the famed aether into the wastepaper basket. In any case, the second axiom forbade light from changing speed, forcing it to disregard the speed of any source that was emitting it.

Yet given that Einstein had shown there was no need for the aether, what in fact was light? All of the wave theories presupposed some propagation medium for it; and the neat electromagnetism equations described its properties and the way it was propagated. However, if light was a wave but there was no aether, then what was oscillating?

It is of course impossible to imagine sound without the existence of air. But Einstein's perceptions not only did away with the existence of some propagation medium for light, they even forbade it. Temporarily, as it would turn out, they deprived light of any remnants of materiality.

As for the actual substance of relativity theory, it has been dogged by misunderstanding right from the outset. There were many who wanted to see arguments in favour of relativism within it. The bombastic phrase "everything is relative" is implied as a general truth in support of naïve comments or philosophical musings. In reality, it would be preferable to call the theory of relativity—a term not coined by Einstein—"the theory of the absolute". And that is because via the theory, the laws of nature acquire absolute significance, regardless of any frame of reference. As Leonard Shlain observes: "The special theory of relativity [...] became a democratic bill of rights for all inertial frames of reference. The theory does not say that everything is relative, but rather that perceptions of the world are observer-dependent. Only light itself, which cannot be used as a platform because nothing of substance can ever attain this speed, can possibly be the ideal—and unattainable—vantage point."[7]

So according to Einstein, light gains supremacy over time and space. That was the innovative break he made with the past. Yet it was a break that led to the perfection of classical theory, not its overthrow. Newton was the God of the Old Testament, and Einstein the representative of the New Testament on Earth.

3.5   The Merging of Time and Space

Readers may well already have come to harbour serious reservations about the great reputation of relativity theory. Being unused to the kind of questions posed by physics, you may be wondering whether that reputation is truly justified, and whether all the fuss about the applicability of physical laws in inertial systems is really worth the effort. The fact that the speed of light remains stable and independent of any motion by the observer or source obviously sounds impressive. Yet neither that nor tossing the aether into the wastepaper basket would suffice to lend special theory the weight it appears to have. It is of course important that the failing unity of physics was restored by the universality of physical laws argued for by relativity. But that is of more interest to those who already have a taste for the philosophical dimensions encompassed by scientific knowledge.

Nevertheless, the edifice of physics has space and time as its indisputable foundations. Both concepts enter into every important area of the subject, be it mechanics, electricity or magnetism and optics. Light's curious property of always being at the same speed, whether we approach its source or draw away from it, could be attributed to its very nature—a whimsical, problematic nature. But Einstein did not take that easy line of thinking. He observed that the concepts of space and time were only weak at those points in optics concerning the speed of light. But since like every velocity, that of light is the quotient of distance over time, he moved on to the audacious hypothesis that the very concepts of both distance—i.e. space—and time were in need of revision. Yet if those notions were to be altered, how would they conform to every other branch of physics, where they had so far seemed to function impeccably? It was as if someone had opened Pandora's box.

The consequences of this reasoning were truly unpredictable, and it was they that lent the theory its special value. First among them was the non-absolute character of time, or its relativity. As a notion it is difficult for the human mind to overcome prejudices and grasp. The truth is that time appears to pass by independently of us, somewhere outside our everyday world; and nothing appears to affect its passing. "And beyond any particular clock," writes A. Lightman, "a vast scaffold of time, stretching across the universe, lays down the law of time equally for all. In this world, a second is a second is a second. Time paces forward with exquisite regularity, at precisely the same velocity in every corner of space. Time is an infinite ruler. Time is absolute."[8]

The absoluteness of time was not only incorporated into philosophical thought, but even into Newton's mathematical worldview. As he notes: "Absolute, true and mathematical time, of itself, and from its own nature, flows equably without relation to anything external…"[9]

All the same, the absoluteness of time is not all that self-evident. And if shaken, it drags down many of the concepts fundamental to science and life—such as that of simultaneity. The need for a new designation of time becomes compelling if we start out from the more familiar notion of space. It has three dimensions—length, width and height—on the basis of which we determine the size of objects. Yet when determining them by the three spatial co-ordinates we refer to the present, or at any rate to some limited interval of time. Space alone and the objects it contains do not mean much without the inclusion of time. Any constituent of the world—be it a cloud, a living organism or our planet itself—exist, function or evolve within a framework known to physicists as the "space–time continuum". Every event in that continuum is characterised by three co-ordinates determining its position in space, and then also a fourth, temporal dimension, giving its location in time.

In reality, this reasoning leads to a kind of merging of time and space. Yet how is such a merger possible, given that we are dealing with entities so very different from one another? Space we can see: we move around in it. But as for time, we are merely aware of its presence. We can neither change it nor stop it. Space is something external; time appears to exist within us. We can grab hold of an object in space. But time—and therein lies our tragedy—is the one grabbing hold of us.

It is thus necessary to stress that as advanced by relativity, space–time is not simply space with another dimension stuck on. In reality, four-dimensional space–time constitutes the underlying fabric of the universe. Prominent mathematician Hermann Minkowski, one of Einstein's professors at Zurich Polytechnic, who, it should be noted, regarded his student as idle and indifferent, was so impressed by relativity that in 1908 he wrote: "Henceforth space by itself, and time by itself, are doomed to fade away into mere shadows, and only a kind of union of the two will preserve an independent reality."[10]

It follows that when light is viewed as a fourth, qualitatively different dimension in space, both geometry and the structure of the universe unfold with great clarity. Then gravity, for instance, is not a force the Earth exerts on other bodies. On the contrary, it is due to the warping of space–time caused by the Earth's presence. From that perspective, gravity resembles the dip created when an iron ball is dropped onto a net.

Contrary to our intuition, the second axiom of relativity now forces us to accept that time does not pass everywhere in the same way. As Einstein drily observes: "An hour sitting with a pretty girl on a park bench passes like a minute, but a minute sitting on a hot stove seems like an hour."[11]

Needless to say, the non-absoluteness of time has nothing to do with experiences of that kind. It is simply an attribute of the universe, a requirement of its profoundest laws. Time is not absolute, but relative to the observer's motion and the event being observed. Two adjacent stationary observers would obviously agree on their measurements of the space and time intervening between given events. Such observers are said to belong to the same region of space–time. But if they are moving relative to each other, each will be in a different space–time region, in which case their measurements of space and time will differ.

It is of course obvious that the new perceptions of time even upset our common sense of temporal sequentiality. Thus, two events not linked by a cause-and-effect relationship may appear in a different order if those observing them are travelling at high speeds. For instance, if someone fires a gun on Earth, and one second later (in her estimation) an astronaut does the same on Mars, a passenger in a high speed rocket may draw the very different conclusion that the gun on Mars was fired first. "In wisdom have You made them all," as the Psalm says—note that if the axiom of relativity precluding the existence of speeds faster than light did not apply, then the sacred principle of causality would be in trouble. In that illogical world, the past would intermingle with the future, and a murder victim might appear to fall to the ground before the killer pulled the trigger. When the theory of relativity began to gain currency, provoking the amazement of the general public, a limerick saying much the same thing was published alongside cartoons, explanations and tongue-in-cheek articles:

> There was a young lady named Bright
>> Whose speed was far faster than light;
>> She set out one day
>> In a relative way
>> And returned on the previous night.[12]

There is thus no universal *now*. Einstein did not simply do away with the notion of the absolute frame of reference. He also overturned the widespread conviction that there was a universal "moment", simultaneous throughout the entire universe. Having studied Einstein's influence on art and literature, Alan Friedman and Carol Donley put it aptly: "The failure of simultaneity to be an absolute property implies that 'the universe at one moment' has no verifiable reality. Moments are not universal; the present is a parochial concept, valid for each observer, but with a different meaning for any observer in any other inertial frame."[13]

It is thus a fact that ever since the notion of time was removed from the pedestal of absoluteness, it has held great surprises in store for humankind. Yet even greater surprises and paradoxes were destined to follow.

3.6   The Age of Light

As we have already said, in the brave new world of relativity, measurements of time and space differ if observers are moving relative to each other. The subtle point, however, lies in the fact that they do not differ at random. According to the new rules, the speed of light must always be the same, regardless of frame of reference and motion. But as we learnt at school, because every speed is the ratio of distance to time, the greater the measured distance in space, the greater the measured time interval must be. Only then will the ratio between the two be ever constant, i.e. equal to the speed of light. Yet the fact that the time interval between two events appears to increase as we speed up leads to the paradoxical conclusion that time dilates!

The dilation of time comes as a jolt to common sense. Its consequences are in any case impressive, being linked to the passing of time, which is such a pressing human concern. The passage of time ceases to be uniform everywhere. If, for instance, this is translated into time as experienced on a clock, the time elapsing between two different readings is greater when the clock is moving than when it remains stationary. In other words, in such cases time dilates and moving clocks run slow. But before readers start wondering how they can still be on time for meetings—if they ever were! —it should be stressed that at everyday speeds, even in a fast car, the time delay is entirely imperceptible.

To avoid simple misunderstandings, we also need to point out that the slow running of a moving clock does not depend on its type or quality, i.e. on whether it is a wonderful old clockwork timepiece or a modern electronic precision clock. Time dilation and slow running occur irrespective of its mechanism. This has nothing to do with the clock itself, but is linked to the deeper texture of time. It is a consequence of the axiomatic premise that the speed of light always remains unchanged. Yet only a genius like Einstein could have grasped that the absoluteness of time had to be overturned in order for that axiom to apply.

At any rate, the conclusion is that the faster a clock is travelling, the slower it seems to run to an observer not accompanying its motion. Nothing out of the ordinary occurs to the moving clock itself. Only its running rate differs, being slower. Thus in a 100 m race, for example, the high precision stop-watches used by the judges might show that the winner covered the distance in 10 s. But the athlete himself, with his even more accurate stopwatch, would discover that he had run the race in 9.999999999999995 s. And being unaware of special relativity, he would quite rightly lodge a complaint with the judges. Of course, such time differences are actually impossible to measure. Even the fastest modern rockets are far too slow for ordinary clocks to gauge the time differences occurring. For the slowing down of time to be visible we would have to go up to speeds that are currently unattainable and may also remain so in the future. For instance, the simple mathematical formulas for relativity show that if a spacecraft were moving at half the speed of light, the second hand on a clock inside it would take fifteen more seconds than a stationary clock to go once round, i.e. to mark the passing of one minute. However, as the spacecraft approached the speed of light, time would dilate considerably. So much so that at around nine tenths of the target, the various events occurring inside the spacecraft would take double the normal time. In other words, they would occur more or less in slow motion. Bearing all of this in mind, the conversation about time in Lewis Carroll's *Alice in Wonderland* does not seem so illogical after all:

'If you knew Time as well as I do,' said the Hatter, 'you wouldn't talk about wasting IT. It's **HIM**.'

'I don't know what you mean,' said Alice.

'Of course you don't!' the Hatter said, tossing his head contemptuously. 'I dare say you never spoke to Time!'

'Perhaps not,' Alice cautiously replied: 'but I know I have to beat time when I learn music.'

'Ah! that accounts for it,' said the Hatter. 'He won't stand beating. Now, if you only kept on good terms with him, he'd do almost anything you liked with the clock.'[14]

However, when the dilation of time is taken to its upper limit, things take on a new dimension. Indeed, a clock passing us by at the speed of light would not seem to be working at all. The intervals corresponding to the movement of the hands would then last an infinity. In other words, time would be frozen. The same applies to a passenger on a train travelling at the speed of light: the platform clock would always show the departure time. In other words, the fortunate or unfortunate passenger would be cut off from the passing of time.

We have seen that clocks or trains travelling at the speed of light cannot possibly exist. Still, we can think of the one and only entity that does have that special ability: light itself. So a clock transported by beams of light would not register any time from the moment light set out to the end of its journey. Even a passenger travelling like the *Arabian Nights* on a special light carpet to magical lands and dales would sense that time was not passing at all—she would remain ageless and immortal.

So however paradoxical it may seem, light has no age, and time does not pass for it. "This view," notes prominent cosmologist Hermann Bondi, "helps to make the unique and universal character of light somewhat clearer. It cannot change once it has been created, owing to the fact that it does not age, and therefore it must remain the same."[15]

One indisputable, earth-shattering truth thus emerges from the theory of relativity. Light does not age, light has no age. It is the sole entity in nature not to be scarred by the passing of time, not to live under its threat. A marvellous Cretan quatrain has grasped this same truth:

> Though times grow ever older
>   and years may pass by too
>   the lamp itself grows older,
>   but its light stays young and true.

Light never grows old, while all the things it surrounds and illuminates—stars and planets, men and machines—suffer the inescapable ravages of time.

3.7   Time Machines and Time-Makers

In all eras, time has been a source of awe and intense concern for human beings. Authors, philosophers and poets have made reference to its inscrutable nature, and have attempted to tame its presence in our feelings and life through words. The unidirectional flow of time, redemption via memory and the sense of transience are highlighted in a unique manner in Marcel Proust's masterpiece *In Search of Lost Time*, while the great poet Paul Celan underlines the hold time has on the things of this world:

> and where they had burnt out,
>   splendid with teats, stood Time
>   on which already grew up
>   and down and away all that
>   is or was or will be -,[16]

Yet though the irreversible flow of time is in absolute control of our existence, and we can neither correct our mistakes nor re-live a wonderful moment, the direction of time does not appear to be of any great importance in science. Its laws apply just as well when time runs backwards. Newtonian mechanics, relativity and quantum physics thus all operate trouble free if the film of a series of events is shown back to front.

Nevertheless, the theory of relativity has played strange tricks with time ever since it did away with its absoluteness, which had seemed sacred and inviolable. What, then, are current views on the age-old human desire to be able to travel in time, to our future or past? It's obvious that we can travel at will in the three dimensions of space—forwards or backwards, left or right, up or down—without much thought, almost automatically, depending only on our needs. But when we get to the fourth dimension, i.e. time, things get tougher: time flows in a straight line, in one dimension, and we flow with it like corks in a river. But if not perhaps interested in their own future, who wouldn't like to know, for instance, what technology will be like in a hundred years' time? Or if it were possible, who wouldn't want to witness the assassination of JFK, which has so many dark sides to it? As early as 1895, H. G. Wells described such journeys in his novel *The Time Machine*. While praising human ingenuity and inventiveness, he also criticized Western civilisation and its self-destructive tendencies.

As for journeying into future time, we have seen that there is a way through: anyone wanting to know what the world will be like in a thousand years just has to travel to a star 500 light years away at 99.999% the speed of light. When she returns, Earth and humanity will be living in the next millennium, but she will only have aged ten years!

The so-called 'twin paradox' likewise rests on the unforeseen dilation of time as introduced by the theory of relativity. Despite being a simple consequence of our new perceptions of time, the paradox sounds like some entertaining science fiction story. The common version involves a space traveller who returns to Earth only to find his twin brother much older than he is. Here is a variation which, while once again underlining the paradoxical things about time, concerns the ancient myth of Ulysses.

After all manner of adventures and great feats, the modern-day Ulysses is living quietly on Ithaca with his faithful Penelope and their son. But being a born explorer, he once again decides to abandon his hearth, this time for the unknown vastness of the universe. He pays no heed to his wife's tears and pleas; yet apart from knowing how to weave skilfully, she is also informed about the theory of relativity, which has just emerged into the limelight. Ulysses boards a spacecraft built for him by a paranoid inventor, and races off almost at the speed of light. Having seen many wondrous things on his travels, he manages to reach an inhabitable planet many light years away.

As a true Ulysses, however, even before commencing his new life he is once again overcome by homesickness for all the good things on Earth he has left behind, for Penelope and for his son, who must now be in adolescence. So he returns in the same spacecraft, having already spent five years of his life as a space traveller. Of course, he is dying to embrace Penelope and see the smoke rising from his hearth. Alas, however, while he is only five years older and wiser—if he is wiser!—Penelope has aged by several decades. Precious few of her features recall her former radiant beauty. As for his young son, he is now a mature man older than Ulysses! All in vain, Penelope swears that the weary but younger traveller who has come from the stars is his very own father. Yet even as regards the rising smoke, the truth is harsh: over the five years the journey has taken, almost half a century has passed on Earth; Ithaca has moved with the times and acquired electric cookers and neon signs.

Travelling into the future thus seems only marginally interesting or encouraging. Things get even worse if we wish to travel backwards in time, to the past, though such a journey has always held a strong attraction for people. Sure enough, there are well-known logical contradictions. For instance, if someone travelling back to the time of the Roman Empire were to notice Brutus' suspicious intentions, they might manage to prevent the assassination of Caesar. But that would change the flow of history, though it has already been written! Nor should the risky side of things be underestimated, since on a journey to the past we might happen to kill one of our ancestors, in which case we would not have been born. One way out of such contradictions is the so-called "alternative histories" hypothesis. On the basis of this, if time travellers ever are led into the past, they enter alternative histories differing from recorded History. In that way they are free to act without being subjected to the coercion of consistency with their own previous history.

At any rate, the truth is that contemporary physics does not shy away from investigating time travel, nor indeed could it, since such investigation is linked to its daring theoretical conceptions. Thus, according to the general theory of relativity, space and time bend. Close to very high-density objects this warping may be almost infinite. A burrow then opens up in space: in other words, a tunnel linking distant areas of space–time as if they were adjacent. This is what the renowned "wormholes" are, by analogy with the holes dug by worms from one end of an apple to the other—physicists are well known for their sense of humour. The famous physicist Kip Thorne and his associates have recently intimated that one such a tunnel would lead to a journey into the past! The problem is that these space burrows have microscopic entrances; and also that they tend to close within fractions of a second.

It's obvious that we are now bordering on science fiction. As to the question of whether we will travel in time over the next century, cosmologist Richard Gott rejoins: "Physicists like me who are investigating time travel are not currently at the point of taking out patents on a time machine. But we are investigating whether building one is possible in principle, under the laws of physics. It's a high-stakes game played by some of the brightest people in the world: Einstein showed that time travel to the future is possible and started the discussion. Kurt Gödel, Kip Thorne, and Stephen Hawking have each been interested in the question of whether time travel to the past is possible. The answer to that question would both give new insights into how the universe works and possibly some clues as to how it began."[17] So the adventure surrounding the meaning of time, which began early on in the last century, shows no sign of ending.

Besides, in his own mythical world Jorge Luis Borges suggests to us that there are other possibilities: "Differing from Newton and Schopenhauer, your ancestor did not think of time as absolute and uniform. He believed in an infinite series of times, in a dizzily growing, ever spreading network of diverging, converging and parallel times. This web of time – the strands of which approach each other, bifurcate, intersect or ignore each other through the centuries – embraces […] *every* possibility. We do not exist in most of them. In some you exist and not I, while in others I do, and you do not, and in yet others both of us exist. In this one, in which chance has favored me, you have come to my gate. In another, you, crossing the garden, have found me dead. In yet another, I say these very same words, but am an error, a phantom."[18]

Great literature and poetry are comforting for the mind and soul. Yet the harsh reality always remains the same. For humans, the passing of time is linear and irreversible. Only light has tamed time in perpetuity.

3.8   The Perilous Journeys of Length and Mass

Time dilation is not the only paradoxical consequence of the axioms of relativity. Bodies moving at very high speeds also are also subject to alteration in shape. This too violates common sense and experience, which hold that the shape of objects is preserved unless deformed by an external force. In our everyday world, a speeding car, a flying bird and even a fast-moving supersonic jet retain shape regardless of their speed and direction.

However, relativity's new perceptions contend otherwise. As objects move in space–time, time measurement is not the only thing to undergo changes. Space does too. And because it is distorted, the shape of objects is also distorted. For instance, from the window of a train travelling at relativistic speed—e.g. half the speed of light—trees or utility poles look thinner. As velocity increases, their upper extremities curve, while rectangular house windows become pointed. It follows that space is not typified by uniformity and inertia, as argued by Euclid. Nor is it absolute, as Newton believed. Instead, it interacts with the shape and size of the objects moving in its domain.

The appearance of objects travelling at relativistic speeds is thus full of surprises: rods bend, and bicycle wheels look like boomerangs. Most characteristically of all, the length of objects appears to contract.

We should note that contraction only occurs in the direction of motion. High precision photography of a jet plane flying at twice the speed of sound would establish that it is fractionally shorter than when on the ground. Of course, the pilot would not feel anything out of the ordinary—to him, the dimensions of the aeroplane and his own self would not change. However, he would see the world around him somewhat compressed, and would feel uncomfortable if unfamiliar with the consequences of relativity. Luckily enough, ordinary planes contract by less than a single atom! But things are entirely different if velocities approach the speed of light. A rocket travelling at four fifths of that speed appears to contract to half of its original length. Should its speed get even closer to that of light, however, the rocket may even contract by nine tenths. Normally speaking, the length should be totally eliminated if the rocket equals the speed of light. This patent contradiction is one of the reasons rendering the speed of light the fastest that does or could exist in nature. A limerick popular among physicists offers an amusing illustration of this paradox:

> An active young fencer named Fisk
>   Had movements exceedingly brisk
>   By the speed of his action
>   Fitzgerald contraction
>   Reduced his épée to a disk

Of course, objects don't actually contract when moving at relativistic speeds. If so, it would be difficult for them to return to their normal state when coming to a standstill. Essentially, contraction illustrates how space is distorted from another frame of reference, just as we measure the distortion of time when establishing that moving clocks run slower. Let us imagine a super-fast relativistic train with just one carriage passing by a station platform. The station manager would see the train shrink in the direction of its motion. But the passengers on board would think the platform was moving, and would conclude that it was contracting rather than the train. Time dilation and length contraction effects are interrelated, and are not due to illusions, they simply arise from the revelatory new view that Einstein brought to our outlook on space and time.

The consequences of this new view do not end here, however. The axiom holding that the speed of light is the fastest in nature also alters our perception of the mass objects have. In our schooldays, we are tediously taught that the mass of an object is constant, and is linked to Newton's laws via force and acceleration. If we push a body, it accelerates; for as long as the thrust remains constant, the body gains ever greater velocity. Its increase in speed cannot be limitless, however, since there is a terminal velocity in the universe, which is the speed of light. And no moving object can accelerate enough to reach that limit, let alone exceed it.

Einstein thus expressed the view that as the speed of an object increases, so does its mass. As a consequence, it yields less and less to the force accelerating it. Indeed, the equation concerned predicts that as the object's speed approaches that of light, its mass comes close to infinity, but then the force required for any further increase in velocity also becomes infinite. The impasse is plain to see, and goes to show that the speed of light really is an unbreakable barrier.

In fact, it is the fastest speed that can exist in nature. Any object moving at a velocity lower than that of light will always stay below that barrier. On the other hand, anything moving at the speed of light must be of zero mass at rest, and is condemned to move at that speed forever. Until recently, it was thought that such a Sisyphean fate was reserved for neutrinos—curious, phantasmal particles which abound throughout the universe. But a sensitive experiment tracking neutrinos at the bottom of a mine proved that they too have infinitesimal mass. So while their speed approaches that of light, they will never attain it. Quantum field theories hold that another category of particles, the gravitons, are the mediators of gravitational force, and like photons supposedly have zero inertial mass. In consequence they too move at the speed of light, though confirming their existence via experiment is far from easy.

3.9   A Paradoxical World Is Confirmed

The consequences of relativity theory—time dilation, length contraction and mass increase—appear so foreign to our own everyday world that one is tempted to regard them as theoretical games. Besides, common sense and insight are shaped by experience, and the laws directly affecting humans are easier to take in. Yet the Earth and its environment make up an insignificant dot in an infinite cosmos. What typifies the universe are vast distances, high speeds and powerful gravitational effects. In such entirely different environments, our own notions of common sense collapse. A form of intelligence capable of crossing the entire universe would form its perceptions of space and time in such a way as to encompass the paradoxical consequences of relativity. Einstein's conception in itself rested on axiomatic and notional principles. That was not only its splendour, but also its power, for it surpassed the everyday world and went in search of substance beyond phenomena. But in the real world, could there possibly be trains or rockets travelling at such high speed that relativity effects become apparent?

Rockets moving at a velocity comparable to light are clearly a far-off and perhaps unattainable technological dream. Yet in the present-day world of large scientific laboratories, relativity is an ineluctable everyday reality. Even in some technological applications such as colour TV, electron speed is reasonably high, and relativity effects have to be borne in mind in appliance design. All the same, one permanent cosmic theatre of such effects, mind-boggling in terms of extent and variety, does exist—that is, the very universe itself! The cataclysmic processes in the stars and galaxies revealed by modern observations can only be interpreted in terms of what both the special and the general theory of relativity contend.

Complex, cutting-edge technology research centres have in any case been built on Earth, in an attempt to partially mimic the extreme energy conditions in the universe. The best known are the European Organization for Nuclear Research (CERN) in Geneva and the now defunct Fermi National Accelerator Laboratory (Fermilab)—named after the great Italian physicist—near Chicago. These centres have accelerators several kilometres in circumference, where subatomic matter particles are sped up with the aid of powerful magnetic fields. They are then made to collide either with each other or with atomic nuclei, allowing valuable conclusions to be drawn regarding the structure of matter and its properties. Inside the accelerators, fundamental matter particles such as electrons and protons gain velocities occasionally exceeding the speed of light. Relativity effects then become apparent. Particle masses increase over a thousandfold, and become more difficult to deflect by magnetic fields. In a ring accelerator at CERN with a radius of 30 km, researchers managed to produce electrons twenty thousand times heavier than an electron at rest. Their speed then reached 0.999999999987 that of the speed of light. To get back to more prosaic numbers, the fact does of course remain that building the accelerator in question cost over one billion dollars. All the same, in the 1990s it lent great impetus to research into the structure of matter.

Nor has the relativity of time remained exclusively in the realm of theory. It is well known—often from bitter experience—that radioactive particles are classified by their lifetime, indicating how long on average they remain stable before decaying. As has been verified, the lifespan of a particle is prolonged with increasing speed. This prolongation is exactly as predicted by Einstein's equations. The human soul would surely very much like its own life to have the same luck!

At any rate, time dilation has been confirmed by specially designed experiments which, rather than making a clock travel at extreme speeds, exploit the fact that modern atomic clocks are incredibly accurate. Scientists put four high precision caesium clocks in passenger jets and sent them around the world twice, once eastwards and once westwards. At the end of the journey, when compared to another atomic clock at the US Naval Observatory, they were found to have lost a few billionths of a second.

The universe itself offers convincing evidence of time dilation, via the cosmic radiation that reaches Earth from the depths of space. This consists of high energy particles such as protons and electrons, which often collide with the oxygen and nitrogen in the upper layers of the atmosphere. At a secondary level this then produces muons, particles 200 times heavier than electrons. The lifetime of muons is so short that they should not reach the Earth, but should decay much earlier. However, because their speed approaches that of light, to an observer located on Earth their life gains in duration. And that permits them to reach the surface of the Earth and be tracked by sensitive instruments. If muons themselves were conscious, they would of course live in their own "normal" time, but would still reach Earth, since their distance from it would be much shorter due to length contraction!

The special theory of relativity is thus no mere theoretical construct that impresses for the neatness of its mathematics and its deeper natural content. Though its influence may not be clearly visible in everyday life, knowledge of it is a sine qua non for the progress of science and modern technology. Indeed, the theory's influence grew in the decades following first publication. Its inevitable merging with the other major theory of the 20th century— quantum physics—not only led to a fuller understanding of the atomic world, but also to significant predictions such as the existence of antimatter, which was triumphantly proved by experiment.

So it is that the special theory of relativity has passed all the experimental tests required of a theory in physics with flying colours. From conjecture and daring predictions, it has now become an unshakeable foundation of our scientific age. As if wishing to persuade even the most incredulous, it has become linked to applications beyond the laboratory and sensitive measurements, thus hallmarking human history and blazing new trails in the era of technology. In fact, an equation that was written within the terms of relativity and accounted for the impossibility of exceeding the speed of light was to presage nuclear energy, its tragic consequences and its beneficial aspects.

3.10 The Equation Typical of Our Times

One of the four scientific papers Einstein published in 1905, while earning his living in the Bern Patent Office, contributed to interpreting a wide range of phenomena, including why the sun shines. It was nevertheless destined to lead to a global equilibrium of fear, and to major peacetime and wartime applications. Einstein's paper concerned the equivalence of mass and energy, and formulated the equation linking the two previously unconnected entities. And though not warranted by its content, the equation really did become common knowledge when the atom bomb unleashed enormous quantities of energy over Hiroshima, sowing destruction.

All the same, it should be stressed from the outset that the famed equation now included in school textbooks and often encountered in science fiction novels and newspaper articles concerned anything but atomic bombs. In a thought experiment related once again to light, Einstein linked and—what's more—equated mass and energy, two entities that had so far appeared to be distinct. In his own words: "The mass of a body is a measure of its energy content."[19] That equivalence is expressed by the equation:

$$E = mc^2$$

which is quite rightly considered typical of our age and its intellectual and technological accomplishments, and also of the dangers inherent in the irrational use of scientific knowledge. In it, Roland Barthes discerned the archetype of revelation: "The historical equation $E = mc^2$, by its unexpected simplicity, almost embodies the pure idea of the key, bare, linear, made of one metal, opening with wholly magic ease a door which had resisted the desperate efforts of centuries."[20]

It's worth noting that the mutual conversion of mass and energy is a common phenomenon, but is not usually perceptible because infinitesimal amounts of energy are involved. When a match is struck, for instance, a chemical reaction converts a very small amount of mass into radiation and the kinetic energy of hot gases. The mass difference that becomes energy is not measurable, however, since it is no more than a billionth of the total!

On the other hand, the coefficient that converts one entity into another is enormous, as it equals the speed of light squared. The energy contained in just one gramme of any material is thus equal to that used daily by the population of a large city. But in contrast to chemical reactions, where energy is derived from the electrons in atoms, nuclear energy comes from artificially splitting the nucleus itself. This is so-called nuclear fission, which then leads to nuclear chain reactions and the release of large quantities of energy. Nuclear energy stands out from other extensive applications of Einstein's equation because the mass differences are substantial. This makes it a highly efficient if scarcely popular source of energy—or destruction. Note that splitting one uranium nucleus releases ten million times more energy than burning one atom of carbon.

Public opinion thus regards the magical relationship $E = mc^2$ as synonymous with nuclear energy. Just like other widespread ideas, though, this is a fallacy. The basic difference lies in the amount of fuel required to produce a given amount of energy in various ways. Even if we were to weigh a coal-fired power station together with the coal and oxygen consumed per week, and then re-weighed it a few days later together with the carbon dioxide and other by-products of combustion, we would once again find that the total weight was somewhat less. So once again in this case mass is converted into energy.

As is obvious, however, the inverse of this equation also applies, meaning that every energy conversion is accompanied by an equivalent increase in mass. For instance, a football gains in mass when kicked by a player's foot; and a glass of milk is of greater mass when the milk is hot. Even winding a clock increases its mass. In all of these cases, though, the energy conversions are minimal and so correspond to incredibly small increases in mass. On the other hand, in large laboratories where microparticles are accelerated to speeds comparable to that of light, electricity consumption is enormous. A proton or electron then acquires a mass many times greater than that it has at rest.

So despite the widespread impression to the contrary, Einstein's famed equation does not refer to atom bombs or nuclear power plants in particular. It simply contributed to the understanding of the associated phenomena, though its significance lies elsewhere. As Stephen Hawking notes, "Some people have blamed the atom bomb on Einstein because he discovered the relationship between mass and energy; but that is like blaming Newton for causing airplanes to crash because he discovered gravity."[21] Yet despite this indisputable truth, humanity has every reason to be concerned: scattered across our planet in states large or small are nuclear arsenals capable of killing or eliminating every life form over large expanses. It has been calculated that the global nuclear stockpile is currently one million times more powerful than the fateful bomb that destroyed Hiroshima!

It is of course true that in recent decades a rough balance had been established between the powerful nuclear states, and both inspections and international treaties were frequent. Circumventions were not unheard of, needless to say, and general concern remained palpable. But the situation has now been worsened by Russia's invasion of Ukraine; along with innocent victims and the flight of refugees, it has unleashed both fears and threats of nuclear interventions.

There can be no doubt that nuclear weapons and, in some cases, nuclear energy itself are a hubris in the sense of Ancient Greek tragedy. Catharsis is thus necessary, but responsibility for it does not only lie with the powerful on this Earth. At the same time, it depends on the knowledge and vigilance of us all. As former nuclear inspector Pantelis Ikonomou astutely concludes: "The scientific world has a historic obligation towards the ultimate goal of completely removing the threat that nuclear weapons pose to the human race. Nuclear disarmament is the mission of saving the world. Yet nuclear disarmament is not an easy process. It cannot be quick, quiet nor cheap. It is immensely complex. But it ought to be done before *the last human error* occurs."[22]

In any event, it is worth underlining that in an equation so impressive for its simplicity and significance, it is the speed of light that serves as the link between energy and mass. The energy conveyed by light itself was what led Einstein to his equation. So yet again, the crucial significance of light in the major strides made by science is clearly apparent.

As unbelievable as it sounds, light itself is often transformed into matter. In the study of cosmic rays, the light-sensitive layer of a photograph first recorded the indisputable conversion of light into a pair of matter particles. The same process was later recorded in amazing photographs depicting the traces of fundamental matter particles in modern accelerators. Analysis of the photographs shows that at some points in space an electron-positron pair is produced, with trajectories that clearly appear to be deflected in a magnetic field. At those points, a photon or light particle—which is electrically neutral and thus leaves no trace—instantaneously converts its energy into matter particles. So the intangible, immaterial entity that is light can be unexpectedly transformed into matter.

As is now believed, similar processes must have been rampant when the universe was in the early stages of creation. Photons, which then abounded, were constantly converted into matter, and vice-versa.

Light, then, has existed since the beginning of the universe—*in the beginning there was Light*—and, having created matter, it reaches us today as witness to the first moments of creation. Another light, that of the sun, is due to nuclear masses being converted into light energy. So as the author now knew, light from the beginning of time, and light from the sun or stars, had always caressed her face and hair. As he continued his aimless wanderings, however, he got the feeling that she had her own ways of remaining bright—perhaps the brightest place in the world.

3.11   God Was a Geometer

As we have seen, the early 20th century was characterised by a genuine revolution in notions of space and time. This revolution was based on a simple question asked by Einstein in his childhood: What would the world look like to a person travelling on a light beam? The answer to that question was provided by the special theory of relativity, which essentially had to rewrite Newtonian mechanics. All the same, Newton's other important conception, concerning the description of gravitational attraction, remained unassailable. In fact, it was so precise that the motions of the planets were entirely predictable and comprehensible.

At some point in the autumn of 1907, however, Einstein had what he himself would describe as the happiest thought of his life. "I was sitting in a chair in the patent office at Bern," he says, "when all of a sudden a thought occurred to me. 'If a person falls freely he will not feel his own weight.' I was startled. This simple thought made a deep impression on me. It impelled me toward a theory of gravitation."[23] Nowadays, with manned spacecraft and the images of astronauts more or less swimming in the void, Einstein's observation seems a commonplace. Yet it took a leap of imagination and a special intuition for its importance to be grasped.

That importance was to be clearly highlighted by the General Theory of Relativity, which Einstein formulated a few years later, in 1916. In real terms, the theory was an entirely new and radical conception of gravity's nature and role.

Gravity is all-pervasive, both in our everyday world and in the farthest corners of the universe, in our solar system as much as in the attraction the Earth exerts on an apple. As Einstein observed, it only disappears if we are in free-fall or moving in orbit around the Earth. In both cases, the acceleration of motion is stable. In an entirely natural way, this finding leads to its inverse: in an enclosed chamber accelerating at a stable rate, the existence of a gravitational field is clearly apparent. For instance, gravitational effects will be apparent in the interior of a rocket moving through interstellar space with the assistance of its engines, since any free object will fall backwards. So what Einstein regarded as a principle—the so-called equivalence principle—was that all bodies respond to a gravitational field in the same way they respond to an accelerating frame of reference. The equivalence principle satisfied Einstein's unshakeable intuition that physical laws must be the same irrespective of the observer's state of motion.

What then was gravity, since it seemed so inextricably bound up with motion type? Here it's worth recalling that there is also another entity every bit as present, and every bit as all-pervasive: space-time. As it emerged from the special theory of relativity, four-dimensional space-time is the basic web on which events unravel. So being inescapably encountered in every corner of space-time, gravity may be connected to its very existence or structure. What served as the link between the two entities was geometry, the very geometry of the universe. But what was needed was to escape the Euclidean paradigm and its wondrous axioms, and be led to another one, capable of corresponding to four-dimensional space-time, which was curved, no less!

It's obviously difficult to imagine curved space, and four-dimensional space at that. Those complex sensory organs we call our eyes are constructed to gather information that only obeys Euclidean geometry. Yet as early on as Dostoevsky's *Brothers Karamazov*, we come across the following amazing comment: "And therefore I declare that I accept God pure and simple. But this, however, needs to be noted: if God exists and if he indeed created the earth, then, as we know perfectly well, he created it in accordance with Euclidean geometry, and he created human reason with a conception of only three dimensions of space. At the same time there were and are even now geometers and philosophers, even some of the most outstanding among them, who doubt that the whole universe, or, even more broadly, the whole of being, was created purely in accordance with Euclidean geometry; they even dare to dream that two parallel lines, which according to Euclid cannot possibly meet on earth, may perhaps meet somewhere in infinity. I, my dear, have come to the conclusion that if I cannot understand even that, then it is not for me to understand about God."[24]

Luckily enough, the impossibility of depicting multi-dimensional curved abstract spaces did not prevent ingenious mathematicians from precisely describing their geometric attributes. One such non-Euclidean geometry, worked out in the main by the German mathematician Riemann, was to form the basis for developing the general theory of relativity. This was no trite matter. "Every step is devilishly difficult,"[25] noted Einstein, who dedicated several years to formulating the theory. But when it was published in 1915, physics was enriched with an incomparable tool for understanding and making cosmological predictions. The universe itself had found its Bible!

The general theory of relativity rests on the revolutionary idea that gravity is not even a force like others. Instead, it is due to the curving of space-time caused by the presence of mass or energy. So the various bodies such as the planets or a satellite do not move in elliptical orbits in space because a "force" called gravity makes them do so. In reality—an incredible reality—they always follow straight courses in a space-time that is nonetheless curved! A straight line is then regarded as the shortest route between two points. On the surface of the Earth, which is a curved two-dimensional space, the straight route—called a *geodesic*, no less—is a great circle. These geodesic curves are followed by aeroplanes, being the shortest path between two airports. By analogy, the same situation applies in space. The mass of the sun, for instance, bends four-dimensional space-time in such a way that a direct route on Earth appears curved in the three-dimensional space familiar to us. A corresponding effect is caused by the shadow of an aeroplane—while the plane's route is a straight line in space, its shadow on the two-dimensional space represented by the ground follows a curved line.

In a non-Euclidean universe, however, one key attribute of space-time is of particular importance: just as the Earth appears flat to someone standing on a sports field, so space-time looks flat on a small scale, and acquires its curvature on the large scale. For as long as we humans live in a curved world but insist it is flat, we are forced to invent forces such as gravity. So in local areas of space-time, special relativity and Euclidean geometry will suffice to describe events; but the same does not apply in expansive areas, where the curvature of space-time becomes apparent.

The important thing is that the degree of curvature of space-time is determined by the presence and distribution of matter itself. If matter in one area of the universe is highly dense, curvature will be correspondingly high.

Consequently, space-time is more warped around the sun than near Earth, which is of much smaller mass. This warping reaches extremes near so-called black holes, where a colossal amount of astral matter is trapped in infinitesimal space.

Gravity thus ceases to be a power of known behaviour but mysterious nature. In Einstein's new universe, a geometric property is simply born in the transition from the flat space-time of everyday life to the curved space-time characteristic of our cosmos. Laws of motion and the physics that dictates them are no longer needed. As John Wheeler aptly sums it up: "Spacetime tells matter how to move; matter tells spacetime how to curve."[26] Effects such as gravitational attraction or the lack of it, as well as the apparent equivalence of those effects to accelerating motion, are simply illusions, created by the constant attempt to follow the "straight lines" of curved space.

Whether in everyday life or in the world of the planets and stars, the picture of gravity that emerges is thus radically transformed. We can imagine space-time like a rubber sheet unfolded and stretched flat. Wherever there is an object, the sheet curves down; and as experience tells us, the depth of the curve depends on the object's mass. The sun, for instance, has the greatest mass in the solar system, and the warp it creates in space-time resembles a deep well in the rubber sheet. As would occur if small balls of differing masses were rolling on the sheet, the routes taken by the planets are determined by the limits of the well surrounding the sun. The same happens with an apple falling to Earth—it is not attracted by some force exerted by the Earth, but is trapped and rolls into the dip the Earth creates in space-time.

So up until recently we had imagined time and space as the stable background or cosmic stage where events took place. With the general theory of relativity things changed, and did so dramatically: space and time are now dynamic quantities that not only influence everything occurring in the universe, but are influenced by it. Consequently, the set for the play acts on the plot itself, but the actors' movements shape the cosmic stage. As for the director, her identity remains unknown. But whatever the case may be, she was a geometer!

3.12   The Bible of the Universe

The basic principle of general relativity that space-time is curved by the presence of matter is clear and simple, though the mathematical equations behind it are complex, and solving them in specific cases is no simple matter. Luckily enough, these complex calculations are not necessary when describing gravity in the phenomena of everyday life. We can get by with Newton's equations and the law of universal gravitation, which are taught ad nauseam in schools—even if few pupils understand their importance! Even the elliptical orbits of the planets in our solar system only differ infinitesimally from those calculated on the basis of new perceptions of gravity.

Yet what finally established relativity theory, catapulting Einstein to fame, once again had to do with light and its mysterious nature. We have seen that one of the consequences of special relativity is the equivalence of mass and energy. They are essentially identical concepts. So if a beam of light possesses energy, it has corresponding mass. As a consequence, it must feel gravitational attraction in the presence of mass. This impalpable pull becomes appreciable when masses are large, as in cases where a light beam passes near a large star; its course should then be curved. In fact, Einstein had calculated the deflection of a beam when light from a distant star happens to pass near the sun.

This prediction was difficult to confirm, however, since the sun's bright light hinders the observation of stars near to it in the sky. But the good Lord did not want doubt to remain for long. So a few years after general relativity had been propounded, a total eclipse of the sun drew astronomers' attention. Since solar light is blocked out by the moon in such cases, the eclipse offered a unique opportunity to test Einstein's paradoxical prediction. The exacting measurements were recorded by the prominent English astronomer Arthur Eddington, who was well acquainted with Einstein's work. "You must be one of three persons in the world who understands relativity," someone allegedly remarked to him.

"I am trying to think who the third person is,"[27] Eddington replied with aplomb, reserving that supreme privilege for himself and Einstein alone.

At any rate, Eddington's measurements from the 1919 solar eclipse did confirm the bending of light—indeed, the results caused quite a stir when announced to a packed auditorium at the Royal Society in London. The eminent English mathematician and philosopher Alfred Whitehead describes the historic moment as follows: "The whole atmosphere of tense interest was exactly that of the Greek drama. We were the chorus commenting on the decree of destiny as disclosed in the development of a supreme incident. There was dramatic quality in the very staging—the traditional ceremonial, and in the background the picture of Newton to remind us that the greatest of scientific generalisations was now, after more than two centuries, to receive its first modification..."[28] As for Einstein, he received the news perfectly calmly. It is said that that when a doctoral student asked how he would have felt if his theory had been proved wrong, he replied: "Then I would have been sorry for the dear Lord; the theory is correct."[29]

The fact is that today, the general theory of relativity is not in any danger whatsoever of being disputed. It has been and is constantly confirmed by astronomical observations and fine-tuned experiments. At the same time, it has led to sensational conclusions about the birth of the universe, the existence of black holes and the evolution of stars.

One other consequence of general theory sounds just as incredible: time passes more slowly near massive bodies! For instance, a clock placed near the sun would run slower than another one that was far away from it. In other words, the stronger the gravitational field, the more the passage of time slows down. This effect assumes impressive dimensions on the edges of a black hole, where gravitational field reaches colossal values, and time even appears to stop.

The gravitational dilation of time can even be verified by terrestrial experiments, provided high precision clocks are used. For example, it was shown that an atomic clock at an altitude of 1,650 m in the United States gains about five billionths of a second per year compared to its counterpart at Greenwich Observatory, which ticks a few dozen metres above sea level. This occurs because the gravitational field gets stronger the closer one approaches the centre of the Earth.

It goes without saying that the gravitational dilation of time, being in many ways analogous to that of special relativity, is in no way related to clock mechanisms. It is the rate of time itself that is affected by a gravitational field. So people living in ground floor flats don't violate any scientific truth if they feel a sense of satisfaction when comparing themselves to the snooty residents in the penthouses above. The latter may well enjoy more light and a good view, but on the ground floor you age more slowly!

The general theory of relativity thus slowly but surely gained currency, thanks both to its theoretical structure and to experimental verification. So the fact that it has flourished in such a unique manner, particularly in recent years, comes as no surprise. Words that fire the imagination of modern people, such as quasars, black holes, neutron stars—even the fate of the universe itself—are rooted in what Einstein's genius discovered. With the general theory of relativity the universe really did gain its own Bible, and its pages have proved both inexhaustible and miraculous.

3.13   The Discovery of the Century

As we have seen, many predictions made by the general theory of relativity were confirmed, rapidly lending it a prestige all of its own. But there was one other prediction that took 100 years to verify—the existence of gravitational waves. Einstein had himself referred to their existence, while at the same time expressing doubt as to the possibility of their being detected.

This doubt ceased to exist thanks to a complex yet ingenious experiment that will go down in history under the acronym LIGO (Laser Interferometer Gravitational-wave Observatory, Fig. 3.1). But what are gravitational waves? By rough analogy, they resemble the ripples caused by a boat in calm water, which spread over an ever-increasing part of its surface. In the same way, according to the General Theory of Relativity, gravitational waves are ripples in the space–time continuum due to the motion of celestial bodies or violent phenomena in the universe. These include mergers of black holes, collisions between neutron stars and cataclysmic supernova explosions. Even two people dancing around each other—in other words, possessing energy and mass—can generate gravitational waves, though in that case the fluctuations in space–time are too weak to be detected.

Fig. 3.1 Ripples in the fabric of spacetime produced by two merging black holes were detected by the Advanced Laser Interferometer Gravitational-Wave Observatory (LIGO; top inset). The Dark Energy Camera at the Cerro Tololo Inter-American Observatory (bottom inset) was used to search for the optical counterpart of the gravitational wave event (SXS, LIGO Lab and T. Abbot and NOAO/AURA/NSF/NoirLab/ CC 4.0)

But let's get back to the Gravitational Wave Observatory, which had the honour of first detecting these waves. LIGO comprises two identical hypersensitive detectors installed roughly 3,000 km apart: one in Louisiana and the other in Washington State. Each detector consists of two 4 km tubes at right angles to one other, forming a giant L. From the outside they look like oil pipelines, but in imitation of the vacuum in space they contain no air. The idea behind the experiment is that as gravitational waves warp space–time as they propagate, they should alter the length of each tube. The length will increase or decrease depending on the wave's direction in space.

The difficulty lies in the fact that the above alteration will be infinitesimal: if we imagine a rod of light ten light years long in millimetres—in other words, 1 followed by twenty zeroes—then the gravitational wave would only change its length by one millimetre!

Obviously, measuring the distortion caused by a gravitational wave in space–time calls for a sensitive, innovative technology. Although describing it is beyond the scope of this book, it's worth underlining that light once again plays the leading role. As can be inferred from the initials in the LIGO experiment acronym, the huge tubes are traversed by laser beams, which are reflected by appropriately placed mirrors. From the interference effects produced, it is possible to detect the change in the length of the apparatus's arms. Any such infinitesimal alteration will signify the presence of gravitational waves arriving at this peculiar observatory from the distant universe.

Oh wonder of wonders! One September morning in the year 2015, the two LIGO observatories simultaneously recorded the passing of a gravitational wave. The signal only lasted a few tenths of a second, starting from low frequencies to reach higher ones. "Sounding more like a fleeting thump," comments physicist and author Marcia Bartusiak, "it was music nonetheless to LIGO scientists' ears, the glissando they had been waiting decades to hear."[30] So it came as no surprise that the first detection of gravitational waves was followed, in 2017, by the award of the Nobel Prize to the pioneers behind the idea: Rainer Weiss, Kip Thorne and Barry Barish.

One mind-boggling fact was the origin of the gravitational waves detected by the LIGO experiment. It was established that they were caused by the collision and merger of two enormous black holes, each of which was roughly thirty times the mass of the sun. But apart from the size of the two dark entities, their distance from our treasured Earth was also impressive, being close on 1.3 million light years away!

Both the difficulty and the scale of the experiment are also obvious from the number of researchers participating: over 1,000 scientists collaborated with LIGO, coming from 15 countries and 83 universities and research centres.

Following the first detection of gravitational waves, by March 2020 ninety cosmic events of a similar nature had been recorded. Of particular interest was the detection of a wave originating in the merger of two neutron stars, the characteristic thing being that it was confirmed by observations on various electromagnetic spectrum wavelengths.

There can be no doubt that the detection of gravitational waves threw open a fascinating new window on studying the universe. From now on, its characteristics will no longer be revealed solely thanks to the major observatories dotted on mountains and plains around our planet. Nor solely by the amazing photographs and data collected by spacecraft expeditions to nearby regions in the universe. Apparently, in future an ever-increasing role will go to the stories told by gravitational waves. "It's the first time the universe has spoken to us through gravitational waves," stressed David Reize, LIGO director and leading specialist in laser spectroscopy. "And, as we open a new window in astronomy, we may see things that we never saw before."[31]

Besides, scientists' optimism knows no bounds, and is often reminiscent of pages from science fiction. Many of them envision gravitational wave detectors placed on spacecraft to avoid any terrestrial interference. Still others believe that even the origin of the universe itself will be understood, since gravitational waves have existed from the beginning, long before light was born in the cosmos. Yet there is one undeniable fact: the discovery of gravitational waves marked the culmination of experiments which, in combination with the development of technology and astrophysics, confirmed the General Theory of Relativity once and for all. That is what dictates our revolutionary views not only on the structure of the universe, but on its very birth and evolution. It is no exaggeration to say that modern cosmology rests on the General Theory of Relativity; in addition to the religious or philosophical versions of the creation and evolution of the universe that humanity has come to know down the course of the centuries, we have now acquired a more powerful one, imbued with the prestige of science.

4

The Quantum Realm of the Microcosm

4.1 The Road to Quantum Theory

Major scientific achievements usually prove durable, it's just that from time to time some partial revision or conceptual extension is called for. So Newtonian mechanics and the laws of universal attraction remain valid, even though both special and general relativity pointed out weaknesses in them under certain conditions, and the need for time, space and gravity to be viewed differently. In a similar manner, a few decades after the wave nature of light and electromagnetic radiation were confirmed, some new experimental data showed up gaps that still existed. A better or different explanation seemed necessary.

The most marked discrepancy between electromagnetic theory and experimentation concerned the radiation emitted by bodies when heated. What was more, the so-called spectral lines typical of light emissions from atoms did not tally with the predictions of classical theory. Unknowingly, physics was slowly but surely leaving the classical domain and entering the quantum realm.

It's worth underlining from the outset that the quantum realm is in essence that of the microcosm. Within its hazy borders, electrons, atomic nuclei and all sorts of small matter particles are composed and interact to make up the wondrous structure of the world as we experience it in everyday life. The paper this book is made of has its roots in that microcosm of matter. As well as being responsible for the light of the stars, it is the foundation on which the amazing technological inventions of our times rely to operate. The laws governing the wondrous realm of the quanta—the paradoxical quantum laws—not only explain why a table does not break down into its atomic

G. Grammatikakis, *The Autobiography of Light*,
https://doi.org/10.1007/978-3-031-56917-3_4

constituents, but also how a complex chemical reaction occurs. If planetary motion is due to gravitational attraction, if the diffusion of electromagnetic waves obeys neat equations, and if relativity reveals the attributes of space–time, then in the quantum realm the structure of matter reigns supreme. The laws governing it make up a new edifice, which was to be named "quantum mechanics"—the mechanics of the microcosm.

Of all the theoretical achievements in physics, perhaps only quantum mechanics can claim to be a true revolution. As Victor Weisskopf notes, "Relativity theory [...] is, in some ways, the crown and synthesis of nineteenth-century physics, rather than a break with the classical tradition. Quantum theory, however, was such a break; it was a step into the unknown, into a world of ideas that did not fit into the web of ideas of nineteenth-century physics. New ways of formulating, of thinking, had to be created in order to gain insight into the world of atoms and molecules..."[1]

That being said, we ought to note that the first step towards the quantum realm was made in 1900, by a major physicist who was nonetheless a conservative thinker, hidebound by classical ideas: one Max Planck. He himself strove—unsuccessfully—to reconcile the old with the new. The step towards the quantum realm was made in an empirical way, so its importance was slow to emerge. In his time as a professor of Physics at the University of Berlin, Planck gave a mathematical description of how thermal energy is emitted when a body such as a lamp filament or red-hot poker is at a high temperature. Yet as he was amazed to realise, one precondition of his empirical formula was the unwelcome admission that radiation was not emitted continuously, as dictated by electromagnetic theory, but in discrete quantities of energy. Planck named these infinitesimal amounts quanta; a word which, in its simple or compound forms, was to hallmark physics over the following decades.

The idea that certain physical quantities existed in discrete amounts was obviously not new, nor did it seem paradoxical to begin with. Matter, for instance, has been known to consist of atoms since the time of Democritus, and so can be regarded as quantised. For instance, the mass of a gold bar is an unknown but true multiple of the mass of a single atom of gold. The same holds for electricity: all electrical charges are a true multiple of the charge in a single electron, so electricity is quantised too. Yet the idea of energy being quantised is radically different from what classical physics decrees. Any system in Newtonian physics, be it a clock pendulum or a falling apple, can have any energy value, though obviously within certain limits. In other words, energy values form a continuous spectrum, like the uninterrupted flow of a liquid. By contrast, quantum physics only allows for certain energy values; nature resembles a cash machine that only dispenses specific banknotes. As the philosopher Bertrand Russell put it: "No adage had seemed more respectable in philosophy than 'natura non facit saltum', Nature makes no jumps. But if there is one thing more than another that the experience of a long life has taught me, it is that Latin tags always express falsehoods; and so it has proved in this case. Apparently Nature does make jumps, not only now and then, but whenever a body emits light, as well as on certain other occasions. The German physicist Planck was the first to demonstrate the necessity of jumps."[2]

So nature did make jumps—but how big was each one? Obviously very small, since in everyday experience, nature and energy never ceased to appear as an uninterrupted continuity. It was difficult to imagine the heat emitted by a red-hot body or the light radiation we get from the sun as being made up of bursts of energy, or bundles too small to perceive.

And yet that was how things were. The truly minuscule "packet" of energy in each quantum was easily ascertainable. All one had to do was multiply the radiation frequency by a new fundamental constant in nature, known as *Planck's constant*, which was represented by the letter h. Note that the h symbol comes from mathematics, where it expresses an infinitesimal difference. Here again, Russell comments, "It is such a small quantity that, except where measurement can reach a very high degree of accuracy, the departure from continuity is not appreciable."[3] The constant's numerical value is extremely small—if written as a decimal number, it has over thirty zeroes after the decimal point.

The h constant is now found in books as well as on commemorative postage stamps, in research papers and on posters. It truly does have an important role to play in the microcosm, as one of the fundamental constants revealing and regulating the world's hidden structure. God may be eternally absent, but He has at least left His fingerprints behind!

Planck was well respected and exceptionally lucky in his scientific career. Born in the German city of Kiel, he grew up in an intellectual environment, as his father was a professor of Constitutional Law. Although he showed great flair for music and literature, he eventually studied Physics in Munich, and soon became a professor at the University of Berlin.

On the other hand, Planck's personal life was unusually tragic. In World War I he lost one of his sons at around the same time both of his daughters died of childbirth complications. His house and invaluable scientific archives were razed to the ground in World War II bombing raids. Thereafter, his second son was accused of plotting against Hitler and executed by the Nazis. While on a walk in the woods near Berlin in his youth, it was he who had heard his father confiding prophetically that his discovery might prove as important as those of Copernicus or Newton.

And it truly did. The only thing was that making its true significance emerge called for a scientific genius less hidebound than Planck by traditional ways of thinking and the limits imposed by classical physics. That genius existed, and had only just begun to earn a name in the scientific community. Yet again, it was none other than Albert Einstein. He was to accept and expand the onrush of quantum ideas into physics, only later to become their greatest opponent.

4.2   Photons Are Here to Stay

Light was struggling to blaze the trail to the quantum world, opening up cracks in the edifice of classical physics. To begin with, those cracks were all but imperceptible. As we have seen, the first of them concerned the light energy emitted by a heated body, which Planck was forced to admit was quantised rather than continuous in form. The second crack appeared when classical perceptions failed to account for a simple phenomenon known as the photoelectric effect.

Light falling on a metal surface is capable of dislodging electrons. In fact, this ability grows stronger as the frequency of the light beam increases, if for example it is in the ultraviolet range. The phenomenon is aptly named the photoelectric effect, since in a closed circuit the electrons leaving the metal cause an electric current to flow. That is why similar setups are now used in light meters and electronic eyes.

Although the photoelectric effect had been known of since the late nineteenth century, it was hard to imagine it providing crucial, conclusive proof of the quantum nature of light radiation. Nevertheless, in one of the papers he published in 1905, Albert Einstein founded his explanation for the effect on a simple yet for him typically revolutionary hypothesis. By it, he pushed Planck's timid quest to its limits: while Planck ascribed the quantisation of thermal energy to atomic oscillation, Einstein argued that every form of electromagnetic radiation consisted of quanta. It was they that made up its deeper texture. Thus, light itself consisted of minute amounts or "quanta" of energy, later to be named *photons*. Only the existence of photons could satisfactorily account for the photoelectric effect. When light falls on metal, each electron absorbs a photon, gaining the energy it requires to be released. So light is a current or rain of photons.

Entitled "On a Heuristic Point of View Concerning the Production and Transformation of Light", this was the paper that won Einstein the Nobel Prize, not relativity. As he himself states: "According to the assumption considered here, in the propagation of a light ray emitted from a point source, the energy is not distributed continuously over ever-increasing volumes of space, but consists of a finite number of energy quanta localized at points of space that move without dividing, and can be absorbed or generated only as complete units."[4]

In short, light has some kind of particulate texture; it is made up of minute "grains", each of which carries a specific amount of energy. Thus, one of the greatest and hardest won certainties in physics—the wave nature of light— was once again beginning to falter. Slowly but surely, Newton was exacting his revenge!

Yet what energy did each of these paradoxical particles have? According to Einstein, who extended Planck's formula on the subject, the energy in each photon is easy to calculate. Once again, it is the famed h constant multiplied by the wavelength of the radiation concerned. But given that the constant's numerical value is exceptionally small, the energy of a photon is also almost inconceivably minuscule. Suffice it to say that if left switched on for 24 hours, an ordinary 100-watt incandescent light bulb emits roughly twenty trillion trillion photons. In other words, 1 followed by 24 zeroes!

According to the basic ratio determining the energy of a photon, because a quantum of ultraviolet light is at a higher frequency, it carries more energy than its infrared equivalent. That is why we tan in the summer, when the ultraviolet light reaching the Earth is stronger. The neat and simple relationship linking light energy to radiation frequency did more than merely account for the photoelectric effect; as was obvious, it contained a revelatory new image of the world. Light clearly also displayed a particulate structure that conflicted with its wave nature. This new attire for light—the quantum attire—amazed the scientific community of the time. Realising the magnitude of the contradiction, Einstein himself was to write: "[…] the next phase in the development of theoretical physics will bring us a theory of light that can be understood as a kind of fusion of the wave and emission theories of light."[5] The interesting thing is that when a constellation of brilliant scientists did begin to shape that theory, which was Quantum Mechanics, Einstein constantly raised objections to it. Yet he himself had opened Pandora's box, and was perhaps scared at what it contained. Towards the end of his life, he wrote to a friend: "The entire 50 years of deliberate pondering have not brought me closer to an answer to the question, 'What are light quanta?'

Nowadays every Tom, Dick and Harry thinks he knows it, but he deceives himself."[6]

All the same, more than a little was already known about photons on the basis of relativity theory. Since their energy was determined by their frequency, they had to have mass. But how did that mass not become infinite, given that they travelled at the speed of light? Here lay the peculiarity of the photon that lent it its unique, prominent position among the pantheon of fundamental particles: the photon was the only particle to have a rest mass of zero. However, on account of its constant motion it acquired mass and fixed energy. It was precisely this mass that resulted in light being bent by gravity, which served as triumphant confirmation of the general theory of relativity.

So gravity acts on a photon, but is incapable of making it move faster or slower. "As nature's most subtle creation," physics professor and author Michael Sobel stresses, "the photon teeters on a knife-edge. With rest mass equal to zero, if it traveled the slightest bit slower than c, its energy and mass would become zero. It would have ceased to exist. Alternatively, if the photon traveled at c but had a rest mass the slightest bit greater than zero, it would have infinite energy and, in effect, could never have been created."[7]

New experiments in around 1922 confirmed the need for photons to exist, and the basic relation determining their energy. This time it was established that when X-rays—i.e. high energy photons—collided with electrons, they behaved like incredibly small billiard balls. Though somewhat reluctantly, few physicists could harbour any further doubts as to the existence of light particles. Of course, that did not mean that light did not also display wave motion. A photon had the attributes of a particle, yet still retained those of a wave. Sometimes it appeared in one form, and sometimes in the other. Duality, one of the most bizarre characteristics of the quantum world, had timidly begun to emerge.

So photons had come into science to stay. In fact, they slowly but surely gained a unique position among matter and energy particles. Yet while they were to continue their triumphant journey, another revolution had begun on a different road: that of the atomic world. Once again, it was a revolution based on light, quantisation and quanta.

4.3   Quantum Leaps in the Atom

As is well known, the fact that matter consists of minute components or atoms was a bold idea put forward by Democritus. This idea may seem somewhat jaded nowadays, when we even talk of living in the atomic age, our children learn the structure of the atom from very early on, and so many technological innovations owe their existence to atomic processes. Yet it has never ceased to be one of the greatest ideas ever conceived by the human mind. Its significance is on a par with the universality of gravitational attraction, or the electromagnetic nature of light, which are however much more recent ideas. In one of his lectures, Richard Feynman asked his audience to imagine a deluge that would destroy all of scientific knowledge. Were it to be saved, which sentence would be capable of containing the most valuable information in the fewest words? It was, he answered, "that all things are made of atoms – little particles that move around in perpetual motion."[8]

It should be noted that Democritus' central concept was not simply that matter comprised fundamental constituents that could not be further divided, and were thus "individuals" (Greek: *atoma*). He also attempted to account for attributes and phenomena such as taste and smell on the basis of the atom's attributes. The path taken by modern physics did not in essence aim at anything more—but this time the path rested on experiment as its final arbiter.

To cut a long story short, in the early decades of the twentieth century it was confirmed that an ordinary atom had a central nucleus consisting of protons. Almost all of the atom's mass is concentrated in that nucleus. At a considerable distance from it, electrons orbit like planets around the sun. The force holding them is electromagnetic in nature, since protons have a positive and electrons a negative charge. And, as often occurs in human relationships, opposites attract.

Hydrogen is the simplest atom, with one proton in its nucleus and one electron in orbit around it. Every time an electron is added to a particular orbit, a new chemical element takes its place, whether large or small, in the universe. Helium, for instance, has two electrons, carbon twelve, oxygen sixteen, and so on. It is hardly strange that the simplest atom—hydrogen—is highly abundant in the universe. And that is because more complex atoms mean more complex nuclei, which are not easy to produce. Note that the colossal Big Bang mechanism that is believed to have created the universe only managed to get as far the helium nucleus! The role of creating more complex nuclei was later taken on by the stars, via the nuclear processes occurring in their interiors.

Investigating the structure of the atom proved to be yet another dramatic adventure. But leaving that aside, let's take a slightly closer look at the orbits of electrons around the nucleus. In a manner of speaking, that is, since the dimensions of the atom are so small that even the most powerful electron microscope can only photograph it as a fuzzy dot. Nor, for that matter, are the "orbits" of electrons true orbits—but for the time being it is instructive to regard them as being so. The important thing is that every element known to nature has a unique, specific arrangement of electron orbits. In oxygen atoms, for example, both the number of electrons and the radius of each orbit are always the same, regardless of whether the atoms are moving freely in the air or forming part of some complex chemical compound. So each of the 92 elements existing in nature has its own, distinct orbit pattern, in some sense its own hallmark.

Here a basic question crops up, the answer to which is key to understanding our world: if electrons are supposedly orbiting around their nucleus, why should atoms be stable? Essentially, electromagnetic theory does not allow this. As the electrons orbit, they should constantly radiate light; light which would in fact belong to a frequency spectrum like the colours of the rainbow. That way electrons would constantly lose energy, and would rapidly spiral down into the nucleus. Having taken up the quantum theory baton from Planck and Einstein, the Danish physicist Niels Bohr wrote: "My starting point was not at all the idea that an atom is a small-scale planetary system and as such governed by the laws of astronomy. I never took things as literally as that. My starting point was rather the stability of matter, a pure miracle when considered from the standpoint of classical physics."[9]

Indeed, Bohr was the first person to make an attempt at accounting for the stability of matter. A dialectical thinker full of personal charm, he was described by Einstein as "an extremely sensitive lad and one who moves around in this world as if in a trance."[10] He had been born into an intellectual environment, and on completing his doctoral dissertation in Copenhagen was to continue his physics research in England. Mainly under the guidance of Ernest Rutherford, the science laboratories there lay at the forefront of research into atomic structure. Once exposed to contemporary scientific thinking, Bohr succeeded in combining the classic image of the atom as a planetary system, which was only accepted by Rutherford's enthusiastic followers, with the quantum ideas proposed by Planck. When he returned to Copenhagen as a professor in 1916, Bohr had already formulated the basic principles of his own version of the atom and its laws.

The cornerstone of Bohr's thought was, once again, the nature of light. He could not possibly accept the particulate theory, which had returned to the limelight with Einstein's photons. Bohr believed that light's incontrovertibly quantum behaviour was not rooted in its own nature, but in the atom itself. So according to his version of things, electrons only moved in firmly fixed orbits around the nucleus. Depending on its orbit, each electron had a specific energy value that hinged on its orbital radius, and since intermediate values could not exist, electron energy was quantised!

Quantised orbits were Bohr's first step in violating the classical framework. Theoretically, a satellite can move in any orbit around the Earth; and the same applies to any planet going round the sun. But the second step, although it resulted from the first, was even more daring: by absorbing the precise amount of energy required, an electron could perform a leap—a quantum leap—to a higher orbit that had an empty space. Or, vice-versa, it could drop down to a lower orbit, emitting the specific energy difference between the two levels in the form of radiation.

In other words, quantum leaps in the atom's interior were not jumps to random orbits. They always had to give or take a "quantum" of energy; and one which, as Planck had correctly supposed, corresponded precisely to the product of the h constant multiplied by the frequency. At no time did the electron have the right to be in a position between two allowed orbits. It jumped from one to the other, without crossing the intermediate space! Although this obviously seems paradoxical, we should remember that the way actors move in films creates the illusion of continuity, while in reality it comes from adding up distinct frames. Note that according to these views, photons only took part in absorbing or emitting light from matter. In a vacuum, light could always be regarded as travelling casually and uninterruptedly as an electromagnetic wave. Quantum behaviour was simply a "memory" of how light was produced, and had nothing to do with its real nature, which was purely oscillatory.

If quantum jumps seem strange to you, no need to worry. The leading physicists of the time reacted in much the same way. When Pauli commented that it was all pure madness, Heisenberg retorted that it was a madness with a method. Meanwhile, in a conversation with Bohr himself, Schrödinger vented his indignation: "If all this damned quantum jumping were really here to stay, I should be sorry I ever got involved with quantum theory."[11] Luckily he did not regret it in practice, since he was later destined to formulate the equation characterising the quantum world.

Experimentally, at any rate, quantum jumps did not yield results in the least bit crazy. Far from it. In its ground state, the atom is stable, since a large energy gap—imagine a step higher than you are tall—prevents the electron from reaching the first excited level. Nor is it able to plummet down either, since there is no available orbit. So although by microcosmic standards there is an enormous gap between the electron and the nucleus, the atom usually behaves like a steel ball. Only high temperatures are capable of exciting it; and as excited electrons return to lower levels, light quanta are produced.

The light of a lamp or a star are thus due to subtle processes in the interior of a large number of atoms.

Although the interior of an atom is obviously invisible, there is a window that lets us peer inside: the atom's spectrum, which can be recorded using special laboratory instruments. Each element has its own characteristic spectrum, which is not of course the same as the colour spectrum revealed by Newton when he analysed white light. Elemental spectra have a number of bright lines characteristic of each element. If an electron jumps between levels with a high energy difference, the photon emitted will be of equally high energy, corresponding to ultraviolet light. Conversely, any emission of red light, which is low frequency, means that the electron drop has covered a smaller energy difference.

Since every element has its own characteristic energy levels, it also has its own spectral line imprint. It is its ID: an electronic ID, no less! Thus the visible spectrum of hydrogen consists of three rather vivid lines -a blue one, a blue-green one and a red one—produced by billions of atoms performing identical jumps. The red line corresponds to the sole electron jumping from the third orbit to the second, and the blue-green one to it jumping from the fourth to the second. Much of our knowledge in astrophysics is precisely due to the spectra of celestial bodies; the characteristic wavelengths of light they emit or absorb permit us to trace the chemical elements they are composed of.

In short, once laid bare by Bohr, the atom constituted an entire, minute and orderly universe. And just like Newton's universe, it too had its own eloquent mathematical depiction, though one based on substantial quantum innovations. As a keen lover of art and poetry, Bohr himself was to write: "When it comes to atoms, language can only be used as in poetry. The poet, too, is not nearly so concerned with describing facts as with creating images."[12]

Niels Bohr was awarded the Nobel Prize for Physics in 1922, for his research into the structure of atoms and the radiation they emit. As the quantum world was emerging, one of Bohr's major initiatives was to create a hive of scientific activity in the beautiful city of Copenhagen. In 1920, the Institute for Theoretical Physics was founded there with the financial support of a world-famous Danish beer; just as with art, science is no stranger to patrons or patronage. As the Institute's director, Bohr was to prove his talent in attracting leading scientists and bringing new ones to the fore.

The institute rapidly gained an international reputation, and students from all over Europe set their sights on getting to Copenhagen, not only for the wealth of scientific ideas circulating there, but also (or mainly!) for the unrivalled ambience of high spirits and intellectual exuberance. The mood was of course set by Bohr himself, who was in the habit of thinking out loud even when playing ping pong, or arguing in pubs about the interpretation of quantum physics. Another thing was that an in-house bulletin analysed all the latest ideas in physics in a comic strip. When someone remarked to Bohr that the institute showed a lack of respect, he not only agreed, but added: "There are things that are so serious that you can only joke about them."[13]

All the same, in all the experiments and debates, disagreements and theoretical constructs, one new element was dominant: that the microcosm of matter was a special realm which, like all realms, had its own laws and logic. Some chinks of light had begun to appear in its seemingly impregnable walls.

4.4   The Dual Nature of Things

Nowadays, with the sobering benefit of time, we can perhaps pinpoint the root of these contradictions. It was the powerful influence still exercised by classical notions on the new ideas that kept emerging in such a dramatic manner. It's worth remembering that the triumph of classical physics rested on segregating the world into two mutually exclusive entities: on the one hand there were particles, which moved in firmly fixed orbits and obeyed the laws of mechanics; and on the other hand there were waves—e.g. electromagnetic waves—which propagated through space carrying energy, and obeying the precise equations of wave motion.

However absolute this distinction seemed to be, the photon hypothesis appeared to make an unexpected breach in it. With its wave nature confirmed, light was simultaneously gaining a particulate one. Could it then be that the classical distinction between waves and particles only existed on the surface, corresponding to our limited knowledge and perceptions? And if a wave such as light really did consist of particles, why could the opposite not also be true? In other words, why should a particulate material not behave as a wave?

This hypothesis obviously sounds overbold. Yet if a particle such as a proton or an electron also led some kind of existence as a wave, then the equilibrium would be restored, and fit in with the symmetry and economy constantly sought for in the natural world. The particle is of course an entirely comprehensible and specific concept, but experience tells us that linking it to some sort of wave is hard to accept. In the brave new world of quantum physics, however, that link was what helped cut the Gordian knot.

The revolutionary idea that every electron belonging to an atom could be associated with a wave was conceived by the French physicist Louis de Broglie, amazingly enough in his doctoral dissertation. He had initially studied history, but during World War One served in a radio station in the Eiffel Tower, where his experience with radio waves diverted his attention to physics. Notably, De Broglie submitted his thesis to the famed Sorbonne. As a duke, he was the scion of a noble and highly influential family, so the university professors found themselves in something of a quandary. On the one hand, accepting bizarre ideas about particles that were simultaneously waves would jeopardise the university's reputation. On the other, rejecting them bore the fear of reprisals due to the candidate's powerful connections. The dilemma was resolved when someone came up with the bright idea of sending the dissertation to Einstein, who opined that the thesis was impressive, even if grounded in apparently crazy ideas.

That was how the way was paved both for the dual nature of particles and for recognition of the young scientist, who was awarded the Nobel Prize in 1929. In fact, it is the only Nobel award ever to have been made for a doctoral dissertation—a poor precedent, perhaps, since from then on de Broglie made no substantial contribution to the advancement of physics. At any rate, the wave nature of the electron was expressed as a simple mathematical relation, linking the electron's wavelength to its velocity. More generally, the higher a particle's speed or mass, the lower its wavelength. Indeed, it was hardly strange that this ratio once again contained Planck's constant.

Soon, very few people questioned the validity of de Broglie's ideas. Laboratory experiments conducted in America and Scotland confirmed the wave nature of the electron beyond doubt. It is both amusing and typical of an adventurous journey that the eminent English physicist J. J. Thomson won the Nobel Prize in 1906 because his measurements had led to the electron being located, and had proved it was a particle. In 1937, his son shared the same prize with C. J. Davisson because the diffraction and interference effects observed by them proved the electron to be a wave!

Arbitrary orbits now acquired some theoretical grounding. Electron waves had to move in an orbit without any phase difference. Quantisation was the consequence. The circumference of the first (innermost) orbit equalled one wavelength, that of the second equalled two, and so on. That was the reason why electrons did not spiral down towards the nucleus, which would result in the atom shrinking: the smallest orbit's circumference could not be less than one wavelength. The idea of waves always accompanying electrons like halos gave a natural solution to the problem of how matter was so miraculously stable. Natural in a manner of speaking, that is; for beyond the bold idea that the electron was simultaneously a wave, the nature of that wave remained obscure. Another unanswered question was when or how an electron decided to make a quantum jump to a lower orbit. Readers are also sure to be wondering whether wave behaviour was exclusive to electrons, or whether it could likewise be detected in other kinds of matter particles, such as protons or neutrons.

As to the last question, the answer is a categorical yes. Wave nature is common to all material bodies! Both very small ones—such as neutrons—and those of greater mass, such as footballs. Wave nature was not a characteristic peculiar to the microcosm, but an attribute of matter itself. "I was convinced," wrote de Broglie, "that the wave-particle duality discovered by Einstein in his theory of light quanta was absolutely general and extended to all of the physical world, and it seemed certain to me, therefore, that the propagation of a wave is associated with the motion of a particle of any sort – photon, electron, proton or any other."[14]

So even bodies in the macrocosm such as planets or human beings have their own characteristic quantum wave. The reason why it is imperceptible lies in the ratio formulated by de Broglie. When a body's mass is large, the corresponding wavelength becomes extremely short. Thus, the moving electrons in home appliances have a wavelength of roughly a millionth of a centimetre, but in a football the equivalent figure is in the order of a decimal point followed by 32 zeroes! In other words, it is infinitesimally small. And in practical terms, the wave accompanying a moving human being is negligible. From one point of view that is just as well. Otherwise, one curious consequence of quantum behaviour known as *tunnelling* would sometimes allow the human body to go through an armchair and fall down onto the ground.

Photons, then, were not the only things to have wave properties. Electrons themselves also showed diffraction and interference effects. The electron microscopes so widely used nowadays are one impressive application of the wave nature of electrons. Objects are viewed through them using beams of electrons, which focus and form images just as light beams do. But because electrons have a much shorter wavelength than visible light, their waves can penetrate matter to reveal substantial details.

If your first meeting with the realm of quantum physics has left you feeling a little strange, that's not your fault! It is hard to accept that light waves, which interfere and diffract like all other waves, behave like quantum particles when interacting with matter; or that electrons, which travel in a straight line and collide like all other matter particles, show diffraction and interference effects like all waves do.

All the same, the apparent confusion caused by wave-particle duality may have a hidden kind of order to it. Bohr tried to bring this to the fore by using the concept of "complementarity", according to which the dual nature of the wave-particle is made up of complementary attributes. Which of the two manifests itself depends on the phenomenon in hand, just like a coin, which may turn up heads or tails, but never both together. Thus in light, the wave and particle attributes complement or mutually exclude each other, yet both are vital to our understanding of it. As J. Jeans writes: "The wave-picture and the particle-picture do not show two different things, but two aspects of the same thing. They are simply partial pictures which are appropriate to different sets of circumstances [..] and so are complementary but not additive. As soon as light shows the properties of particles, its wave properties disappear, and vice versa; the two sets of properties are never in evidence at the same time."[15]

The idea that apparent opposites go to make up a whole is obviously not new. It was prevalent in the ancient civilisations of the East, incorporated in their myths and philosophy. The famous circular yin-yang symbol of interlocking dark and light symbolises precisely the fact that opposites complement each other. Wherever there is yin, or one side of the circle, there is also yang, or the other. Wherever there is night there is also day; wherever there is death there is life. It is hardly surprising that when Niels Bohr was knighted in 1947, he chose the yin-yang symbol for his coat of arms.

At any rate, as the eminent historian Eric Hobsbawm astutely observes: "Bohr's 'complementarity' was not intended to advance the research of the atomic scientists, but rather to comfort them by justifying their confusions. Its appeal lies outside the field of reason. For while we all, and not least intelligent scientists, know that there are different ways of perceiving the same reality, sometimes non-comparable or even contradictory, but all needed to grasp it in its totality, we still have no idea how we connect them."[16] That very same question has tormented the author of this book, especially in his hours of solitude. He has grappled in vain to conceive of her in her entirety, wishing to comprehend her expression and movements. Yet in the end, whether out of exhaustion or knowledge, he has simply yielded to her charm.

Regardless of philosophical or conceptual reservations, however, the experimental and theoretical strides made by quantum physics were already indisputable. A new world had arisen, though one that strove constantly to hide its true face and significance.

4.5   God Does Play Dice with the World

As it turned out, the microcosm of the atom and electron had character-istics all of its own: wave behaviour and quantisation, orbits that were not orbits but stationary waves, and a dual yet not contradictory nature. In other words, particles were waves at one and the same time. In contrast to the logic behind the macrocosm, where Newton's laws ruled supreme, the quantum world appeared to be full of paradoxes and new realities. Could it possibly have its own inviolable laws? And were they capable of predicting how the atomic world behaved as accurately as their Newtonian counterparts, and of expressing its paradoxes in mathematical terms?

The answer was mind-blowing: yes, they could, but on the basis of proba-bilities! The strict determinism of Newtonian mechanics was to be supplanted by the random and the vague. In identical experiments, similar electrons could behave differently! "God does not play dice with the universe," Einstein retorted, when quantum probabilism began to gain ground apace. In fact, Einstein spent the remainder of his scientific career doggedly attempting to restore the deterministic world's shaken prestige. In vain! God really did play dice with the universe.

The new order became apparent in 1926, when Erwin Schrödinger, an Austrian physicist and professor at the University of Berlin, wrote the equa-tion describing the quantum world. This he did almost overnight. Impressed by de Broglie's work, he had thought long and hard about wave-particle ideas. But when he was invited to give a lecture on the subject, the ideas he presented came under fire for being incomplete. To examine the waves in full, he was told, he would need the equation for them. It is said that early the following morning Schrödinger returned triumphant. "Here it is," he said, writing his now famous equation on the board.

In terms of importance, although the Schrödinger equation is to quantum mechanics roughly what Newton's equations are to the mechanics of the macrocosm, it differs substantially in its mathematical structure. Deep down it is a wave equation, since it had to be in line with what had been discovered so far. A mysterious mathematical function represented by the Greek letter Ψ (psi) is the key to this. As a function, it has an extraordinary ability to depict the atom in all its states. Thus the electron in a hydrogen atom may not only be found in one of its fixed orbits, but anywhere from the atomic nucleus to infinity, though there is a specific probability of it being in a given location at some point in time. This probability is calculated on the basis of the psi-function, which is the heart and soul of the Schrödinger equation.

So this is where the first essential difference or true break from classical perceptions arises. In the atomic world, no-one can know the point in an atom where the electron is located; all that can be known is the probability of it being there. The orbit assumes the form of a cloud—a probability cloud surrounding the nucleus.

It should be noted that Schrödinger himself reacted against the wave being interpreted as a probability wave. It was an interpretation inspired and established by Max Born, another leading quantum mechanics researcher. Indeed, in the same framework he even predicted quantum jumps between the discrete states of the electron. At any rate, this does away with the logical contradiction between waves simultaneously being particles, or vice-versa.

Readers may quite rightly be feeling a little lost by now. First there was Bohr, with electron orbits and how they were quantised. Then de Broglie also predicted something later proved by experiment: that electrons were simultaneously waves, with their stationary states surrounding the nucleus. Lastly, Schrödinger formulated a unique equation with solutions corresponding to the various states of the atom, and precisely predicted the probability of each particular state occurring. The orbit began to yield to a strange "cloud"—not a true one, but a probability cloud. In other words, a non-material, tasteless and odourless cloud, yet one useful nonetheless, since it indicated where the electron was most likely to be encountered. And indeed, lo and behold, calculations showed that there really was almost zero probability of the electron being where early quantum theories, using somewhat arbitrary arguments, had forbidden it from being.

If you haven't run out of patience, you have just witnessed the continuous drawing closer to truth that is a process common in science. This was gained by steps, sometimes sure-footed and sometimes stumbling, by interpreting individual phenomena, synthesizing and revising. Like it or not, in the end the physicists of the time (Fig. 4.1) were persuaded that the quantum microcosm was governed by unexpected mechanics, the basis of which was a wave equation. Previous attempts had not been entirely wrong, but were merely partial depictions of a complex reality.

Nowadays, quantum mechanics is the bedrock of atomic and molecular physics: it interprets the nuclear reactions in stars or light emission from atoms, and to a great extent guides physicochemical processes. Though its role is often hard to discern, quantum mechanics and its laws are the driving force behind a host of applications that have revolutionized our everyday life, from the smartphones that have become an extension of our hand to semiconductors or GPS navigation systems, and from the omnipotent presence of

Fig. 4.1 The Fifth Solvay Conference on Physics ("Electrons and Photons"), 24–29 October 1927, was attended by almost all of those who contributed to the development of quantum theory. Seventeen of the 29 attendees were or later became Nobel Prize winners. Top row (left to right): A. Piccard, E. Henriot, P. Ehrenfest, E. Herzen, Th. De Donder, E. Schrödinger, J. E. Verschaeffelt, W. Pauli, W. Heisenberg, R. H. Fowler, L. Brillouin; middle row: P. Debye, M. Knudsen, W. L. Bragg, H. A. Kramers, P. A. M. Dirac, A. H. Compton, L. de Broglie, M. Born, N. Bohr; front row: I. Langmuir, M. Planck, M. Curie, H. A. Lorentz, A. Einstein, P. Langevin, Ch. E. Guye, C. T. R. Wilson, O. W. Richardson (Benjamin Couprie and Institut International de Physique de Solvay/PD)

lasers encountered when using credit cards and in medicine to the quantum computers that are still in their infancy, but offer significant prospects.

So it will be no exaggeration if our era, accompanied as it is by various monikers, is eventually regarded as the era of quantum mechanics. On its arduous course it has unlocked the secrets of the microcosm and equipped humanity with limitless powers. Of course, as history teaches, these powers do not always act for our good. Knowledge of the secrets of the nucleus, for instance, led to nuclear weapons and their nightmarish presence on our planet. On the other hand, there seems to be no end to the beneficial uses of technology rooted in quantum mechanics and its absurd rules. One need only think of the developments in diagnostic medicine that have saved human lives, or the major particle accelerators that have expanded our knowledge of the structure of matter and the very evolution of the universe.

In sum, the quantum realm does not have fixed borders. It holds ever looser sway as we pass from the atomic process governed by the Planck constant to the brave world of our everyday life. Here, gravity remains the undisputed overlord. At this point we should note that unifying general relativity theory with quantum mechanics—science's two major conquests—remains an abiding dream in physics. Yet nature jealously guards one of its great secrets, perhaps to remind humans of their limits.

4.6   The Certainty of Ignorance

The probabilistic world of quantum mechanics is thus an indisputable reality. But what characterises it is one unexpected, bitter constraint: a particle's motion, i.e. its velocity and position, can never be precisely determined. The more we try to learn its position, the less we know of its velocity. The same applies to energy and time, another equally important pair. The more we try to accurately determine a particle's energy change, the less accurate we are in determining the time that change lasted.

There is, then, a fundamental limitation on knowledge of the physical world. This limitation is known as the uncertainty or indeterminacy principle. This important concept was formulated by the brilliant German physicist Werner Heisenberg, who was only twenty-six years old at the time.

Heisenberg belongs to the multitude of great physicists who shaped quantum theory in the 1920s, one of the most productive decades ever seen in science. Born in a small Bavarian town in 1901, he studied at the universities of Munich and Göttingen. While still very young he became a professor at the University of Leipzig, and was later director of the famed Max Planck Institute for research. Heisenberg was well versed in the classics—his father was a distinguished professor of Ancient Greek—and the influence of Platonic ideas is clearly visible in both his philosophical and scientific thought. "The elementary particles in Plato's *Timaeus*," he writes, "are finally not substance but mathematical forms. 'All things are numbers,' is a sentence attributed to Pythagoras. The only mathematical forms available at that time were such geometric forms as the regular solids or the triangles which form their surface. In modern quantum theory there can be no doubt that the elementary particles will finally also be mathematical forms..."[17]

Before Schrödinger formulated his own more intelligible version, Heisenberg had used austere mathematical language to work out a comprehensive yet esoteric mechanics of the quantum world; although this set out from different principles, in practical terms it ended in the same subversive results. For his contribution to quantum theory and its application to the hydrogen atom, Heisenberg was awarded a Nobel Prize in 1932. His ideas and philosophical personality had a profound, wide-ranging effect.

The uncertainty principle (or more precisely *indeterminacy* principle) which bears his name has consequences far more interesting than first impressions suggest. As we have seen, the principle forbids us from fully knowing both the position and the velocity of a quantum particle simultaneously. It should be stressed from the outset that this is not due to some flaw in scientific instruments that will be eliminated once they are perfected. It is an innate, profounder restriction on the operation of nature itself; a restriction yet again determined by the Planck constant.

At any rate, the first and possibly greatest consequence arising from the uncertainty principle is the stability of matter. "A pure miracle when considered from the standpoint of classical physics,"[18] as Bohr had earlier stressed. Because although atoms are more or less hollow, on the inside they remain undisturbed and timeless—with their electrons ever moving around their nucleus—despite constant collisions and alterations. Bohr had already offered an initial explanation for the stability of matter, but the completion of quantum theory was to lend final meaning to this wondrous characteristic of our world. Sure enough: the uncertainty principle equips the electron with a limitless ability to withstand nuclear attraction and avoid falling into the nucleus, which would be disastrous for the very structure of the world—because that would mean its position could be located, and thus its energy would tend towards infinity. But since the electron, being a particle-wave, also has the tendency to spread throughout space, the uncertainty principle imposes a necessary compromise. The electron moves at an intermediary, stable distance, always aiming to have minimal total energy. As paradoxical as it may sound, then, there is an uncertainty hiding behind the certain, stable structure of our world.

It goes without saying that Newtonian mechanics does not appear to know of such uncertainties. There, the magnitudes of motion can be measured accurately; errors regarding velocity or position are due to imperfections in instruments, and may either be eliminated by repeating measurements or calculated using simple mathematics. So it is no surprise that in the late nineteenth century the universe was regarded as a well-wound clock. From the moment it was set in motion—for whatever reason—its operation was entirely predictable, at least in theory. Faith in this was put succinctly by Laplace, a leading astronomer and mathematician. As he wrote: "Assume an intelligence that at a given moment knows all the forces that animate nature as well as the momentary positions of all things of which the universe consists, and further that it is sufficiently powerful to perform a calculation based on these data. It would then include in the same formulation the motions of the largest bodies in the universe and those of the smallest atoms. To it, nothing would be uncertain. Both future and past would be present before its eyes."[19]

The classical universe was thus disciplined and predictable. Almost tedious. Every future event was the result of causes at some preceding moment. Luckily enough, a universe of that kind, which glorified determinism and the power of prediction, began to collapse under the indeterminacy principle and quantum physics in general.

We should note that just as relativity theory was misinterpreted to create the impression that everything—moral values included! —is relative, so the uncertainty principle reinforced many people's belief that everything is uncertain. And though that may well be true of the current political and economic climate, it has nothing whatsoever to do with the uncertainty principle. References to it when commenting on social or even artistic phenomena quite rightly led a reliable magazine to formulate the so-called quantum theory of journalism, according to which whenever journalists or commentators invoke the uncertainty principle to justify the unjustifiable, there is a strong probability they have no idea what they are talking about.

At any rate, it is worth underlining that the indeterminacy necessarily governing our knowledge of atomic processes is rooted in light, the very medium we use to perceive the world. In its quantum nature, no less. It is as if God has given humans a valuable gift, but knowing their insatiable mind, He has seen to it that the gift simultaneously sets their limits. And so it is: even with the very best experimental apparatus, observing an electron's motion necessarily involves illuminating it. But being a particle, a photon also carries energy. So illuminating a particle inevitably disturbs it to some degree.

Things now end up in a vicious circle. If we want to determine the particle's position accurately, light of the shortest possible wavelength must be used. So by reducing the wavelength as much as we can, we inevitably increase the frequency of the photons, i.e. their energy. But then, although we have gained something as regards determining the position, we have lost out when it comes to determining the particle's velocity. The greater the photon's energy, the greater the disturbance to the motion of the electron we are optimistically trying to measure.

It's a true impasse. Let's try doing the opposite: causing minimal disturbance to the system, so as to get a truly accurate measurement of its velocity. Once again, according to the basic relationship linking energy and frequency, this means using very low frequency light. We've lost again, as a small frequency means a long wavelength. And that, in turn, implies increased inde-terminacy in measuring position! In essence, then, moving from low to high wavelength quanta only shunts the difficulty around, but does not do away with it.

But if that's really how things are, why are we right in claiming that for everyday phenomena, such as the motion of a car or a football, our measure-ments are as accurate as possible? One way or the other, light is obviously used to take the necessary measurements. Could it then be that the indeterminacy principle does not apply in the macrocosm?

We have already seen several times—in relativity and quantum mechanics—that no strict distinction of that kind could exist, as it would put paid to the highly-prized unity of nature. In simple terms, indeterminacy has negligible consequences in our world. As already mentioned, that is because the value of Planck's constant, which dictates the limits of our knowledge, is exceptionally small. If it were not so, a game of billiards would be reminis-cent of the quantum experience had by one of George Gamow's characters: "Watching the rolling ball, Mr. Tompkins noticed to his great surprise that the ball began to 'spread out'. This was the only expression he could find for the strange behaviour of the ball which, moving across the green field, seemed to become more and more washed out, losing its sharp contours. It looked as if not one ball was rolling across the table but a great number of balls, all partially penetrating into each other. Mr Tompkins had often observed anal-ogous phenomena before, but today he had not taken a single drop of whisky and he could not understand why it was happening now."[20]

One of the boldest consequences imposed by the indeterminacy principle is thus that the notion of an orbit comes apart, since any observation of an electron that could contribute to determining its motion disturbs that very motion to an unacceptable degree. In other words, quantum physics forces us to accept that while a particle does move, it does not have an orbit!

In this way, the atom is rendered an ever more paradoxical reality. While supposedly being the building block of the material word, every recognisable trait of it vanishes in the twinkling of an eye. Its electrons do not move in a true orbit, nor is their position precisely known at any given moment in time. Yet with their discontinuous quantum jumps, electrons emit light. Lastly, the nature of atomic particles has dual form: as a wave and as a particle. Neverthe-less, no-one can determine which of these two states an electron or a proton is in. Any measurement ruins the accuracy of either their wave or their particle properties!

There can be no doubt, then, that quantum mechanics overturned the thinking on which science had rested for centuries. Yet it also ran contrary to human experience: it spoke of orbits dissolving into probability clouds; it believed that particles were waves at one and the same time; and it argued that measurement alone highlighted one of the two potential states coexisting in a body until then.

True chaos? Anything but. It was just that we had to change our perspective of the world. As time moved on, no doubts remained that quantum mechanics offered a correct theory of the microcosm. Indeed, equipped with the indeterminacy principle, it acquired an impressive completeness and solved specific problems, while fresh experimental data repeatedly confirmed its predictive capability. Besides, the applications that were to create today's technoculture in a few short decades had already begun to appear on the horizon.

Quantum mechanics is thus the grammar and syntax in the theatrical production of the universe: a drama in which Einstein is the set designer, Newton the stage manager and Maxwell the lighting technician. The director, if one does exist, remains unknown.

4.7   Quantum Acrobatics

Quantum mechanics not only broke down the world of experience, but also the solid world of causality and Newtonian mechanics. The fact that its brief was restricted to the microcosm was precious little consolation. In the final analysis, what Nobel Laureate Odysseas Elytis referred to as "this small world, the great" was the fundamental level of matter; its importance in comprehending entirety was, de facto, of crucial significance.

While no-one now doubts the validity of quantum theory, the philosophical and epistemological problems it posed remain largely unresolved. Nevertheless, the debates that often reached fever pitch or ended in delicate intellectual acrobatics have died down. The probabilistic, non-deterministic interpretation of quantum mechanics had the ever-active Bohr as its chief proponent, along with a host of major physicists who had contributed to its formation. As the renowned Copenhagen School, their views rapidly gained considerable influence. In the opposite camp—though with significant differences between them—were Albert Einstein, de Broglie, Schrödinger, Heisenberg and others, who acknowledged the theory's computational power, but balked at its lack of determinism. The dispute, possibly one of the most serious in the history of science, lasted for close on four decades, during which the defenders of determinism lost ever more ground.

According to the Copenhagen interpretation, which soon became quantum orthodoxy, describing an object could not be distinguished from the process of observing it. As the Nobel Prize-winning physicist Steven Weinberg comments: "Where human beings had no special status in Newtonian physics, in the Copenhagen interpretation of quantum mechanics humans play an essential role in giving meaning to the wave function by the act of measurement. And where the Newtonian physicist spoke of precise predictions the quantum mechanician now offers only calculations of probabilities, thus seeming to make room for human free will or divine intervention."[21]

In one extreme version shared by Bohr himself, the Copenhagen interpretation believed that an object's existence only ever acquired meaning when observed. "There is no quantum world. There is only an abstract quantum physical description,"[22] Bohr says. This bold conclusion may plunge some philosophers into the depths of despair, while leading others to exult, but one science journalist showed more aplomb: "When you look at your reflection in the bathroom mirror every morning, you could be doing yourself a favour. After all, some physicists believe that the most fundamental aspects of the universe do not really exist until they are observed. So you could argue that getting up and stumbling to the bathroom is vital to your well-being."[23]

Opponents of the Copenhagen School did not of course maintain a unified stance. Their more general epistemological position could be termed realistic, since it defended determinism and the existence of a real world but found the complementarity principle hard to stomach. They regarded any existing conceptual difficulties as transient. As Schrödinger wrote: "A widely accepted school of thought maintains that an objective picture of reality—in any of that term—cannot exist at all. Only the optimists among us (and I consider myself one of them) look upon this as a philosophical extravagance born of despair in the face of a grave crisis."[24]

Summing up the situation in more moderate terms, one could say that the problem of measurement emerges in quantum physics, though without any unanimously acceptable solution. The observer and the observed world become entangled in an inscrutable manner never visible to classical physics. In the latter, the world is "out there" and the observer "here". In quantum physics, by contrast, the observer is regarded as more or less essential—in a physical rather than merely philosophical way—if the notion of external reality is to have any significance.

At any rate, without wanting to dive deep into the troubled waters of the dispute, by way of illustration we should mention one example that gave physicists and philosophers a serious headache. It concerns a cat! This animal, known to epistemology as Schrödinger's cat, finds itself in a peculiar predicament: it is shut in a dark box, in which there is a hammer and a flask of poison. The hammer is connected to a radioactive material, but will only fall on the flask and release the poison if radioactive emissions are detected. The question is this: at any given moment, is the cat dead or alive?

Readers may well be concerned about the mental state of physicists who preoccupy themselves with problems like this. Yet the question truly is profound. And that is because radioactive emissions are a purely quantum mechanical phenomenon, and nobody knows the precise moment in time when a particle will leave the nucleus. After a given interval there is only an equal probability that decay may or may not have occurred, i.e. either for the hammer to have broken the flask and the cat to have been sacrificed in the name of science, or for it still to be alive.

Nevertheless, according to the mathematical structure of quantum mechanics when applied consistently to the entire system, there is only one way out of the dilemma. Before we open the box to see what has happened, the cat is simultaneously dead and alive. In other words, it is a spectre, a half-alive, half-dead hybrid. And only our observation will put it in one of those worlds once and for all. Just as the electron is neither a wave nor a particle, and only measurement decides on its state, so the unfortunate pussycat: it is neither dead nor alive, until someone—clearly an animal lover—opens the box. "This some philosophers of science find very hard to accept," observes Stephen Hawking (whose cat is killed by a gun rather than poison). "The cat can't be half shot and half not-shot, they claim, any more than one can be half pregnant. Their difficulty arises because they are implicitly using a classical concept of reality in which an object has a definite single history."[25]

The heated philosophical and epistemological debates that went hand in hand with the shaping of quantum mechanics unfortunately featured the physicists themselves far more than philosophers. At any rate, over time discussion gave way to calm, almost fatalistic acceptance of the theory. The problem of interpretation was left for the future, or for people with the time and inclination for similar worthless intellectual exercises. Not that outbursts were entirely unheard of from time to time. Leading modern thinker Karl Popper was to denounce what he saw as the brainwashing of physicists with the orthodox interpretation of quantum mechanics: "Perhaps the most important point in this brainwashing is *the problem of comprehension*. Part of the teachings of Bohr—mentor and professor to Heisenberg, Pauli and almost all the other founders of quantum mechanics—consisted in the following: 'Do not seek to understand quantum mechanics, it is almost entirely *incomprehensible*.' Bohr attempted to *explain* this incomprehensibility, i.e. to render incomprehensibility comprehensible. His explanation was that we can only understand situations resembling those we are familiar with…"[26]

Yet regardless of its philosophical aspects, it should yet again be underlined that quantum physics, as fully elaborated through the persistent work of ingenious physicists, is a tool of incredible interpretative capability and unquestionable power. One way or another, it enters into almost every modern-day technological and scientific achievement, from the explosion of a star to radioactivity, and from how television works to mobile phones. In practice, then, the prevalent attitude nowadays seems cynical at first sight: while quantum mechanics is used on an everyday basis in several branches of science, concern surrounding the fundamental problem of its interpretation is on the wane. Could this be wise, or is it merely symptomatic of a utilitarian age? "I think I can safely say that nobody understands quantum mechanics,"[27] was how Richard Feynman wryly put it, even though he himself, as one of the giants of twentieth century physics, successfully extended the reach of the subject by describing the interaction of light and matter.

In any case, it is characteristic that quantum mechanics emerged from the need to interpret certain manifestations of light. And while it obviously did not unlock the secret once and for all, it did make a decisive contribution both to understanding light's nature and to the advancement of optics.

4.8   The Photon Love Bond

While physicists paid ever less attention to the problem of interpreting quantum mechanics, enchanted as they were by its completeness and predictive power, one watchful and ingenious mind never ceased to brood over the heresies brought into our solid world by the new theory. It was that of Albert Einstein. As we have seen, having blazed the trail to the paradoxical world of quantum ideas, he rapidly became their most vehement critic. Such a stance is, if truths be told, somewhat unusual in science, but not all that rare in the course of history. Great pioneers often feel the need to distance themselves from their initial radicalism and show a conservative face, perhaps because at some point they discover the consequences of their ideas exceed their intentions, to say nothing of their own selves. Einstein may have found himself overcome by similar sentiments, for in reality he was the last towering figure of a classical era, and could not become a mere disciple of a new one. Whatever his internal motivation, his stance revealed a tragic, perhaps bitter grandeur.

Even if belatedly, Einstein found himself forced to accept that quantum mechanics was both theoretically consistent and strongly backed up by experiment. So instead of disputing it outright, he chose a second line of defence, arguing that quantum mechanics only constituted a partial picture of a far more complex and subtle reality, which remained hidden despite physicists' attempts. In some sense, that reality reflected a more fundamental theory that respected causality and classical ideas in physics. Consequently, quantum oddities should not be seen as so impressive. Uncertain, probabilistic predictions are always the rule when knowledge is fragmentary, in gambling as much as in weather forecasting. Ignorance leads to chance. Science, on the other hand, has always been founded on clarity and logic. Being no more than a partial depiction of true theory, quantum mechanics was incapable of meeting those serious demands.

Whereas other critics of quantum theory restricted themselves to theoretical disquisitions, Einstein was well aware of just how important experiments were. He thus went a crucial step further: in 1935, he and associates Boris Podolsky and Nathan Rosen described a thought experiment designed to show up the weaknesses in quantum mechanics. Named after the initials of its authors, it has gone down in the history of science as the EPR Paradox. Note that yet again, it had to do with the deeper nature of light!

The backbone of this thought experiment lay in the behaviour of two particles that are produced together, but then divide. One typical case is a pair of photons emitted by an atom, each of which takes its own path. Quantum mechanics predicted that these minute light particles would continue to interact instantaneously even if they were millions of kilometres apart, which was, to say the least, incomprehensible. Even if propagated at the speed of light, the interaction would take some time. According to Einstein and his associates, this highlighted an insuperable contradiction in quantum mechanics. Were a convincing experiment ever to be carried out, the truth would shine forth, helping to reveal the true theory of the microcosm.

The way to that revelation was paved by John Bell, an Irish physicist already distinguished for his research into fundamental particles, as well as for a large part of his career spent at the European Organization for Nuclear Research, known to most as CERN. Bell never lost interest in the probabilistic nature of quantum mechanics, which had prompted Einstein to declare his objection in the well-known quip "God does not play dice with the universe." In 1964, Bell was to formulate a theoretical criterion based on the EPR thought experiment that seemed capable of rendering a verdict. It was to go down in the history of physics and ideas more generally as the "Bell inequalities", and its contribution to definitively resolving the debate was to prove decisive.

As Stefanos Trachanas, director of the top-ranking Mathesis MOOC learning platform, lucidly comments: "The issue at stake was of course the debate over the nature of quantum probabilities. On the one hand was the Copenhagen School, which said that quantum probabilities were fundamental (not due to incomplete knowledge); and then there was the "opposition" (or what remained of it after the days of glory), insisting that in the final analysis there were hidden attributes (or hidden variables) to particles in the microcosm, and that the probabilistic behaviour observed in nature was due to our ignorance of those attributes."[28]

At all events, it took close on two decades after the Bell inequalities were posited for the first compelling answer to the dilemma to come. In 1982, a brilliant French physicist named Alain Aspect designed and conducted an experiment that was a clever version of the EPR thought experiment. Since the results turned out to violate the Bell inequalities, the verdict was clear: the sacred principle of causality did not apply in the realm of quantum mechanics, while the amazing "entanglement" of photons revealed by the EPR experiment was an inherent attribute of the quantum world. With Aspect's experiment, the Copenhagen School—shorthand for the group of eminent physicists supporting the probabilistic interpretation of quantum mechanics—scored its first victory.

For obvious reasons, however, there then followed a series of ingenious and sensitive experiments, each addressing the weaknesses of the previous one. Since all of them violated the Bell inequalities, the victory became a triumph. It was thus an act of justice when the pioneers of those experiments were awarded the Nobel Prize for Physics in 2022. They were the French physicist Alain Aspect (obviously!), the American John Clauser and the Austrian Anton Zeilinger. According to the award citation, the scientists were honoured "for experiments with entangled photons, establishing the violation of Bell inequalities and pioneering quantum information science."[29] As is clear from the press release, the value of the experiments prompted by the Bell inequalities does not simply lie in the conclusive, irrefutable confirmation that "God does play dice with the universe". At the same time, emphasis is placed on the emergence of a new kind of technology, which will in future lead to the development of high-performance quantum computers and other applications that are difficult to predict at present.

If anywhere, all of this belongs in the distant future. What matters for the time being is the incontrovertible proof that light has yet another sensational side to it: two photons that have had a common beginning always maintain a covert relationship. They feel for each other and interact regardless of the distance between them. It is as if they were joined by an unbreakable love bond, an unbounded means of communication. Before anything else, this profound secret and its significance call for silent reflection.

5

The Lustrous Dream of Unification

5.1 An Age-Old Quest

Synthesising and unifying apparently unconnected phenomena has always been the main pursuit of physics. More than any other science, physics likes asking about the fundamentals, and abhors complexity. From as early as classical antiquity Democritus had set the tone. With the atom hypothesis he aimed to account for the properties of matter on the basis of a more general unifying understanding. A similar quest runs through the work of both Aristotle and Plato.

The tendency towards synthesizing several phenomena into just a few theories clearly emerged after the seventeenth century, at the time modern science was born. The idea that physical laws were universal is usually attributed to Galileo, who proved that shadow formation obeyed the same rules on the Moon and the Earth; similar ideas appear to have been developed by the Persian scholar and polymath Al-Biruni, an inhabitant of what is now Afghanistan, who wrote treatises on a range of subjects roughly a thousand years ago.

As the breadth of data and discoveries increased, the quest for some unifying scheme remained fervent. A first leap was made by Newton. On the basis of the laws of motion and universal gravitation, he produced the tool for understanding radically different phenomena, from the tides to the trajectory of projectiles, and from the planetary orbits to the way a pencil falls to the ground. His optimism was thus not unwarranted when he wrote in the preface to the *Principia*: "I wish we could derive the rest of the phenomena of Nature by the same kind of reasoning from mechanical principles, for I am

© The Author(s), under exclusive license to Springer Nature
Switzerland AG 2024
G. Grammatikakis, *The Autobiography of Light*,
https://doi.org/10.1007/978-3-031-56917-3_5

induced by many reasons to suspect that they may all depend upon certain forces…"[1]

A second, momentous leap towards a unifying understanding of the world had its beginnings in the experiments carried out by Faraday and Ampère. Despite their different manifestations, electricity and magnetism turned out to be two sides of the same coin—and a valuable coin at that! This was none other than Maxwell's theory of electromagnetism, which embraced electricity, magnetism and light in a common framework. The propagation of light, as well as the deflection of a compass needle, electric current and induction effects were contained in just a few elegant equations, whose predictions were triumphantly verified.

The early decades of the twentieth century were marked by similar achievements. Einstein's scientific quests furthered a wealth of unifying ideas. Thus, the special theory of relativity places time and space on an equal footing, proving that mass and energy are two aspects of the same entity. Meanwhile, with general relativity Einstein interpreted gravity as the warping of space and time caused by the presence of matter. In essence, that is, a fundamental force of nature was equated with geometrical concepts. Besides, Einstein himself strove in vain to the end of his life to formulate a comprehensive unified theory.

The same time saw the emergence of the charming yet provocative quantum world. As quantum mechanics became fully fleshed out, it once again seemed that few phenomena remained uninterpreted. But then the question naturally arose as to whether quantum mechanics was a stand-alone, unique concept, or whether it too could be encompassed within a broader framework. And given that the world turned out to be quantified, what impact did that have on the other grand, classically oriented theories such as electromagnetism and relativity, and vice-versa?

It is worth noting that while physics appears to have been the prime mover in the quest for a unified interpretation, other branches of knowledge did not lag far behind. Freud, for instance, attributed the entire range of human behaviour to the operation of the subconscious, while Marx regarded class struggle as the driving force of history. In parallel, polytheism gave way to faith in and worship of a one and only Supreme Being. It is thus difficult to gauge whether the quest for synthesis reflects a metaphysical or philosophical need, or is simply imposed by the course of science itself.

What remains beyond doubt is that towards the mid-twentieth century, physics appeared to rest on the firm foundations provided by the general theory of relativity and quantum mechanics. Yet the edifice of the cosmos appeared disjointed, since the two theories were in essence irreconcilable. A wall seemed to stand between the microcosm and the macrocosm, as if they were two separate kingdoms with different rulers and laws. Indeed, the ruler of the microcosm—quantum mechanics—was entirely devoid of common sense. She claimed that every matter particle was simultaneously a wave, while with the uncertainty principle she blocked off knowledge of a physical process. So the clash ran deeper, unsettling established views and having undeniable philosophical implications. Consequently, it is hardly strange that to the end of his days Einstein refused to accept the element of chance and unpredictability that quantum mechanics brought to the hitherto deterministic world. Many years later, in response to the great man's stubborn conviction that God did not play dice with the universe, Stephen Hawking replied: "Consideration of black holes suggests, not only that God does play dice, but that he sometimes confuses us by throwing them where they can't be seen."[2]

But the chasm opened up by quantum physics was not the only one on the road to a unified theory. As was soon confirmed, two other forces beyond electromagnetism and gravity had an essential role to play in physical processes: weak and strong atomic force. Were they too independent, standalone entities, or yet again part of some unified function that was destined to be revealed?

Despite the major difficulties encountered from time to time, the dream of a unified synthesis or Theory of Everything has never ceased to exist. Possibly because, more than experimental confirmation or mathematical consistency, it is linked to an internal human desire to understand the world around us in a comprehensive way. In mankind's long history, that is why the dream has assumed different forms: it was initially expressed in myths and gods of the sea and skies, and later via religious systems and faiths. Since our own era is dominated by science, it has been charged with making the dream come true. Indeed, in envisioning the unity of all knowledge, leading biologist Edmund Wilson stresses: "I had experienced the Ionian Enchantment. That recently coined expression I borrow from the physicist and historian Gerald Holton. It means a belief in the unity of the sciences - a conviction, far deeper than a mere working proposition, that the world is orderly and can be explained by a small number of laws. Its roots go back to Thales of Miletus, in Ionia, in the sixth century B.C. [...] In modern physics its focus has been the unification of all forces of nature - electroweak, strong, and gravitation."[3]

Yet before physics could arrive at the unification of all the forces it had discovered, if possible, it had first to incorporate special relativity into quantum mechanics, and then to build bridges—if they existed—between electromagnetism and the paradoxical quantum world. The road to unification resembled a tortuous ascent up a mountain. As the summit drew closer, the view opened out and the landscape acquired new vistas; very soon, a beacon on the summit was to light the way and its difficult passes.

5.2   The Unanticipated Road to Antimatter

When obstacles appear insurmountable, be it in science or the lives of nations, the role of personalities or gifted individuals who can bring about radical change seems to be a *sine qua non*. We tend to forget that nowadays, since in the name of an often-affected progressivism we prefer to attribute the achievements of outstanding individuals to vague groups. Fortunately, this tendency to level down has not gained much ground over the historical course of science. So although it is often said, not entirely wrongly, that things were ripe and that someone else would have discovered relativity, the road would have been long and perhaps impassable without Einstein's catalytic thinking. Netwons, Bohrs and Faradays are far from easy to produce; strangely enough, they seem to be born when the need for them arises. The same is true of Paul Dirac, an ingenious physicist who was destined to give substantial responses to the new demands for unification.

The first demand was for quantum mechanics to be synthesized with the special theory of relativity. The truth was that quantum theory as developed by Schrödinger, Heisenberg and others did not obey the principles of relativity established by Einstein. Dirac succeeded in formulating quantum theory in such a way as to incorporate the fundamental ideas on the wave nature of electrons and the relationship between mass and energy into a unified mathematical edifice. If Einstein with his contempt for quantum theory can be regarded as the last giant of classical physics, then P. A. Dirac—as he always signed himself—was the pioneer of its modern era. So it was only right that in 1933 he was to share the Nobel prize with Schrödinger.

It is worth noting that Dirac, who was born in Bristol in 1902, originally studied engineering. This taught him the value of approximation methods, and confidence in forecasting when it came to physical problems. All the same, he sought to approach the fundamental laws by relying more on intuition that on actual data. As Feynman, his continuer, later commented, "Dirac got his answers by… guessing an equation."[4]

Even in physics, aesthetic values were highly important to Dirac. I myself was fortunate enough to hear him, by then a myth, at a lecture in London, and remember timidly asking him why he believed an equation he had just presented was correct. "Because it's beautiful," he replied. Indeed, it is said that his advice to his postgraduate students in 1974, a few years after moving to the United States, was both bold and risky: he said would prefer them to be more interested in how beautiful their equations were than in what they meant! Perhaps he had been convinced by his compatriot poet John Keats, in the closing lines of his *Ode on a Grecian Urn*:

Beauty is truth, truth beauty—that is all
Ye know on earth, and all ye need to know

Identifying truth with beauty seems to be what nature prescribes, too. Thus while Dirac's efforts were focused on the problems of unification, a new, unanticipated world emerged from his equations: the world of antimatter. Quite literally. His theory linking quantum mechanics to relativity presupposed that in addition to the electron, there also had to be another particle of precisely the same mass but with the opposite charge!

This prediction barely persuaded anyone. Yet a few years later, the American physicist Carl Anderson detected a mysterious particle in the cosmic rays bombarding our Earth from outer space. It had the same mass as an electron but, unexpectedly, was positively charged. It was the first antimatter particle, which was named the positron precisely on account of its charge. Dirac's daring theory had been verified, with Anderson winning the Nobel Prize for Physics in 1936 as its first consequence.

Yet that first antimatter particle was set to gain many companions. Sure enough, after World War II other types of antiparticles began to be discovered in cosmic rays. Their presence later became commonplace in the accelerators built to study the structure of matter, thus confirming an important principle that was simply a natural consequence of Dirac's equations: namely, that every matter particle had a mirror image, which differed only in terms of electric charge and certain magnetic properties. The natural world acquired a deeper symmetry, in which antimatter corresponded to matter. Powerful beams of antiprotons or positrons have even been created and made to collide with their opposites. The annihilation resulting from this creates new particles and light radiation, and yields important information on the structure of matter.

The question then arises as to whether antimatter is encountered anywhere on its own in the universe. In his Nobel Prize acceptance speech, Dirac went so far as to hypothesize that the fact the Earth consisted of matter was probably a coincidence. On some stars the opposite could well apply: antimatter could be predominant. In other words, anti-stars, anti-planets and perhaps even anti-humans could exist.

Before readers hasten to point out that the presence of anti-humans is all too obvious on Earth, it is worthwhile clarifying the concept of anti-matter. For science it does not carry the mystery and emotional charge it does for ordinary people. It is an entirely logical and respectable form of matter, roughly equivalent to its mirror image. The difference mainly lies in opposite electric charges. Given that isolated antiparticles have been observed experimentally, the existence of a type of antimatter with them as building blocks cannot be ruled out. For instance, if an antielectron—i.e. a positron—orbits an antiproton, it forms an atom of antihydrogen. As with ordinary matter, the addition of further positrons forms ever more complex atoms; in such cases the nucleus consists of corresponding antiprotons or antineutrons. So the existence of heavier elements—an entire world of antimatter, as imagined by Dirac—does not violate any of the principles of physics.

The only problem is that such a world is difficult to detect. Its features would not differ in the slightest from those of the ordinary world. A star of antimatter would emit light by precisely the same processes, and would live and evolve in the same way as an ordinary star. Nor would gravitational effects differ: an anti-planet would orbit a hypothetical anti-sun in the same manner as our own planets.

There is, however, one attribute of antimatter that makes the quest for it fruitless, and possibly rather foolhardy. It is that every collision between matter and antimatter, or even mere contact between them, leads to true annihilation. Particles disappear, while copious energy is released in the form of radiation. In other words, matter is converted into light! So the coexistence of matter and antimatter is not possible either in the innumerable stars in the sky or on the planets. The upshot would be a violent, destructive explosion.

All the same, the existence of large quantities of antimatter eternally separated from matter cannot be ruled out. But until (and if ever) a way of potentially detecting it is discovered, there seems to be some basis in the advice a well-known physicist has to give: if fate ever leads to us meeting extraterrestrials, we shouldn't just casually offer them our hand. They might be made of antimatter. Then a friendly handshake would end in a cataclysmic explosion.

In any event, the solution to the enigma of antimatter's potential absence from the planets and stars, i.e. from the entire universe, may lie elsewhere. By that I mean in the modern theories that are evolving very rapidly, striving to understand the creation of the universe from a Big Bang. It may be that the embryonic conditions of the Explosion are where to look for the reasons why particles initially gained the upper hand, and then went on to form atoms rather than antiatoms, making up the matter of today's universe.

But the important thing, at least for our own circumscribed and difficult life, is not that; it's that the theoretical prediction of antimatter that was so triumphantly confirmed rested on principles of symmetry and beauty. So just as in life, the quest for beauty in science reveals an entire world that is trying to whisper the truth. This truth seemed to be known to the enigmatic woman who accompanied the author on his wanderings. Perhaps that's why she often appeared to rejoice in light, yet at other times to be covered in a darkness painful and unjust.

5.3   An Acronym Full of Meaning

Interactions between light and matter lie at the root of a whole host of phenomena, which are at times quantum and at times classical in form, and thus comprise a fundamental key to understanding the world. Indeed, that may be why the amazing theory describing them goes by the weighty, imposing name of Quantum Electrodynamics. In English it is known as QED, recalling the Latin phrase Quod Erat Demonstrandum ("This completes the proof", used by mathematicians). As the name indicates, the theory in hand represents the ultimate synthesis of quantum theory and electromagnetism, making it an acronym full of meaning.

As we have seen, Dirac made steps towards this synthesis by describing the motion of the electron in such a way that it obeyed the special theory of relativity. This first form of quantum electrodynamics was then to be elaborated on by large numbers of leading physicists. Yet while the theory now agreed with a wealth of experimental data, its inadequacies soon became apparent. As Steven Weinberg observes: "For forty years general relativity was widely accepted as the correct theory of gravitation despite the slimness of the evidence for it, because the theory was compellingly beautiful. On the other hand quantum electrodynamics was very early supported by a great wealth of experimental data but nevertheless was viewed with distrust because of an internal theoretical contradiction it seemed could only be resolved in ugly ways."[5]

The task of overcoming this contradiction and putting together a comprehensive theory was accomplished by the ingenious American Richard Feynman, one of the most outstanding physicists in modern times. It should be noted that up until the mid-twentieth century, major developments in science were rooted in European countries. But as Europe suffered moral and financial collapse in the wake of World War II, and a large number of scientists had migrated to the new world, scientific and technological innovation were also to move there. The United States thus became a leader in science just as in everything else—sometimes for the benefit of mankind, but often to its detriment.

The idea that Feynman relied on for his quantum interpretation of electromagnetic effects was as simple as it was daring. He considered that charged particles such as electrons or positrons interacted not because electromagnetic force was somehow magically propagated at a distance, but because between themselves, particles emitted and absorbed light quanta, i.e. photons! It was via them that the electromagnetic force dominating phenomena in the atomic world was propagated. Nevertheless, these photons were phantasmal or virtual, so to speak, and could not be perceived in any way whatsoever. Their actual presence would violate the sacred tenets of energy and momentum conservation, which schoolchildren toil over and often come to hate from high school onwards.

The basic idea that one or more photons are exchanged between two electrons is depicted in simple sketches named Feynman diagrams. Their major significance lies in the fact that in a manner of speaking they indicate the calculations required for each particular process. Yet beyond this, the diagrams symbolize the glorification of the unifying—and to some extent poetic—image of physics.

To non-specialists, a Feynman diagram looks like hieroglyphics, or an ideogram with a weighty and secret meaning. Physicists are more down to earth: in the diagrams they can see the path of an electron emitting a photon, which is itself then absorbed by a neighbouring electron. Particle communication is thus due to the "message" conveyed by the photon, though observers conclude that the charges on particles are what cause attraction or repulsion between them. The message received by one electron on approaching another consists of the command: "I am here, please move." Classical electromagnetism used to believe that the message was conveyed by electromagnetic waves, but from now on that major role is assumed by photons. According to the uncertainty principle, electron orbits are not fixed, nor are photon emission and absorption times precise. So it should come as no surprise that the message bearers or photons lead a fleeting, ephemeral existence.

In practice things are more complex or, from another perspective, more fascinating. And that is because not just one, but two, three or even more photons can be exchanged between electrons or other charged particles. Each of these cases has its respective diagram. The more photons a process needs to transmit the electromagnetic message, the less likely it is. So each diagram is not just a graphic way of describing the process; at the same time it also allows us to calculate the probability of that process occurring. Perhaps nature wants to show us that for each of its levels revealed to humans, it exacts a price in delicate calculations.

There are thus grounds for suspecting that quantum electrodynamics is a mind game in physics, which is of no more interest than for its internal rules. And yet as far as predictions and agreement with experimental data are concerned, it is the most accurate theory ever known to science. For instance, the magnetic field of an electron can be calculated to ten decimal places, which is like measuring the distance from Athens to London roughly to within the width of a human hair!

As elegant as it is accurate, given suitable mathematical skill quantum elec-trodynamics is capable of describing an enormous variety of phenomena. In reality, it can describe all the phenomena in the physical world except for gravity and radioactive decay. Its wide range of applications is due to the fact that such phenomena are rooted in atomic scale interactions, atomic and molecular cohesion or the never-ending game between light and matter. It follows that the combustion of petrol, the hardness of steel, the behaviour of positrons, the magnetic field of electric current and the formation of salt are all subsumed in processes describable by quantum electrodynamics.

Of course, an enormous number of electrons are involved in most of the phenomena familiar to us. In such cases the human brain proves too weak to grasp the complexity of their interactions with photons. Yet the principles of quantum electrodynamics do not appear to be the slightest bit flawed; even when there are few electrons, all experimental checks so far have agreed one hundred percent with the predictions. As Feynman observed: "... while I am describing to you *how* Nature works, you won't understand *why* Nature works that way. But you see, nobody understands that. I can't explain why Nature behaves in this peculiar way."[6]

So it could be that the mind of God is structured like an unimaginably powerful computer. But perhaps because humans are known for their conceit, they have only been permitted to make simple calculations, and merely draw close to the truth.

5.4   The Choreographer of Light

We have seen that calculations in quantum electrodynamics are based on simple diagrams depicting the interactions between electrons and photons. In other words, they resemble alluring choreographies of light in its eternal encounters with matter. The choreographer was Richard Feynman, who, like Newton and Einstein, made a decisive contribution to our understanding of light. It stands to reason that anyone writing a book on the subject is bound to make more detailed reference than usual to what made the man and his life so special.

Richard Feynman was born in a suburb of New York in 1918. With the aid of his father—a lively and inquisitive if not particularly well-educated man— he realised that science was not about memorisation, but about seeking basic principles. As he recalls being told: "You can know the name of a bird in all the languages of the world, but when you're finished, you'll know absolutely nothing whatever about the bird. [...] So let's look at the bird and see what it's *doing* - that's what counts."[7]

As is normally the case with gifted minds in the USA, Feynman gained a solid education at the country's famous universities: initially at MIT, before going on to do a Ph.D.—and publish his first papers—at Princeton University. It was there that he gave his first lecture, to an audience including Einstein, Pauli and the leading mathematician John von Neumann. Like other major physicists of his time, Feynman overcame moral dilemmas and worked for a time on the atomic bomb programme. After the war he was appointed professor at Cornell University and, having taught for a year in Brazil, in 1950 he moved to the equally famous California Institute of Technology (Caltech). In contrast to the previous generation of leading European physicists, Feynman was not particularly cultured, nor was he of a philosophical bent. He represented the aggressive new American pragmatism, and was a bohemian, insatiably curious character with a unique sense of humour. Talking of curiosity, in the entertaining pages of his autobiography he says: "That's a puzzle drive. It's what accounts for my wanting to decipher Mayan hieroglyphics, for trying to open safes."[8]

Feynman's dislike of the abstract expressions in quantum mechanics initially led him to his own graphic formulation. Each quantum process was described as the totality of all possible routes, the sum of all possible stories linking a particle's two different states. These ideas were the forerunner of quantum electrodynamics. And while QED theory stands as his greatest contribution, deservedly winning him the Nobel Prize for Physics in 1965, with his amazing insight and knowledge Feynman also excelled in all manner of research topics, from the study of liquid helium to the internal structure of protons; and from weak atomic forces to a comprehensive theory of positrons, antiparticles which he interpreted as electrons moving backwards in time! "There were, it was said, only two ways of solving difficult problems in physics," wrote one reviewer. "One was to use mathematics; the other was to ask Feynman."[9] Meanwhile, his biographer James Gleick notes: "At twenty-three he was a few years shy of the time when his vision would sweep hawklike across the breadth of physics, but there may now have been no physicist on earth who could match his exuberant command over the native materials of theoretical science."[10]

He was equally exuberant in life. Ever open to challenges, capable of being absorbed by an everyday problem or a woman's charms, he was as frank about his virtues as he was about his vices. He discussed problems in physics with all the leading lights of the age—Bohr, Dirac and Wheeler—but also got to know the famous high roller Nick the Greek, and played drum accompaniments to dance performances. He formulated the most accurate theory in physics, but also won a bet that he could crack a safe full of atomic secrets.

Feynman's contribution to teaching physics was as subversive as his research presence. In his university lectures and numerous popular science talks, problems are treated with a beginner's eye, the emphasis being on precise concepts and key questions. The famed *Feynman Lectures on Physics* recording these lessons in book form have been translated and circulated all over the world. To this day, they still teach students—and often mature scientists, too—to avoid formalism and look for the principles underlying phenomena.

Time now for a personal digression. In 1980 Feynman was the commanding presence at a theoretical physics conference traditionally held in the Orthodox Academy at Kolymbari on Crete. As a member of the conference organizing committee, I was lucky enough to get to know his unusual personality up close. In various ways I experienced the breadth of his knowledge and his peculiar and somewhat earthy charm, as well as his enquiring mind. In fact, I soon began to feel incapable of responding to his volley of questions, concerning anything from a roadside sign to the dogmas of Greek Orthodoxy and on to the Phaistos Disc. Though already suffering from ill health, Feynman dominated the conference proceedings from the front row. He didn't miss a single paper, and his often ironic comments betrayed a genius on constant alert. At his own request he gave three lectures to students who had rushed down from Athens; to make it look respectable, the audience was even boosted by the uninitiated local youth. With a twinkle in his eye, Feynman began by stressing that in the next few hours he would interpret every single physical phenomenon. He was not exaggerating, for the classes were on quantum electrodynamics, and the blackboard rapidly filled with strange symbols like ideograms. Nervy and constantly on the move, Feynman was a far cry from the classic image of a teacher with calm, structured speech. He gave the feeling that he himself was once again living some great adventure, and calling his students to join in. He truly was a choreographer of light.

Feynman's quirky integrity and ability to tackle any kind of problem were to come to the fore in an impressive manner shortly before his death. In 1986, the Challenger space shuttle exploded in the air almost immediately after launching, before the eyes of millions of viewers. US President Ronald Reagan personally called for an in-depth investigation into the tragic accident; NASA, whose public image had been severely dented, somewhat unwisely called on Feynman to take part in the committee. He not only managed to pinpoint the true cause of the accident—the failure of a rubber O-ring at low temperature—but revealed it right in front of an amazed US television public by performing a simple experiment. Nor could he resist a few digs at the system and way NASA was run.

Feynman remained a professor at Caltech until his death, following a lengthy illness, in 1986. The world had not only lost one of its finest theoretical physicists, but a spirit of rare virtues.

His departure from this earthly world really did leave an irreplaceable gap. Fortunately, the inspired choreographer of light did at least leave his choreographies behind him. And they had already begun to guide physics along the difficult road that would perhaps unite all interactions.

5.5   Forces and Weaknesses

Grounded in the exchange of photons between charged particles, quantum electrodynamics succeeded in interpreting the complex phenomena in the interaction of light and matter. Yet these phenomena are not the only ones that go to make up physical reality. Across the entire expanse of the universe, from the deepest level of matter (where nuclei and elementary particles are dominant) to stellar processes, an elaborate world is constantly evolving and changing face. It resembles some enigmatic drama without visible beginning or end, yet which employs thousands of actors and a magnificent stage presence. Who then are its prompters? In other words, which are the forces that dictate its everyday, almost eternal plot?

Strange as it may sound and despite the complexity of phenomena, no more than four physical forces shape our world! This really is striking, since everyday life experience alone would tend to convince us of the opposite. Could the motion of a car, the deflection of a compass needle, complex chemical and radioactive processes, biological phenomena, the rise and fall of the tides and the light of the stars possibly all be rooted in the interaction of just a handful of forces?

Nevertheless, this reductive picture is an article of faith for modern science. The forces that have acted on our world since the far distant past really are only four. In fact, humans do not even indirectly perceive the one that just happens to be the strongest, even if it is the one that binds protons and neutrons tightly to the nucleus, lending matter its stability.

But let's begin with sights familiar from experience. The best known, most familiar of the forces is obviously gravity. It is a manifestation of the universal attraction acting between all material bodies. Its laws were given accurate mathematical expression by Newton, permitting us to describe an apple falling to the ground, the motions of the planets or the orbit of a comet passing through our solar system. These laws are not complex: everyone can (or should) remember from school that attraction between bodies rapidly decreases as the distance between them increases, that it is directly proportional to the bodies' masses, and that it increases when masses are large.

So gravitational force is ubiquitous. Its magnitude can become infinitesimal if the masses are small or the distance between them is great. Even astronauts who appear to be swimming in the void are subject to gravitational effects, albeit weak ones. Characteristically, gravitational attraction on the surface of the moon is about a sixth of that on Earth for the same body, so it's hardly strange that the astronauts there walk in leaps and bounds.

Although Newton left no gaps in his mathematical depiction of gravity, analysing its origin and deeper relationship with space took many decades. As we have seen, the relationship was revealed by the grand, solid edifice known as the general theory of relativity.

Apart from gravity, electromagnetism—the second force in this brief survey—is also familiar to humans. With its assistance, electrons are bound around nuclei to form atoms, and atoms are bound to each other when forming molecules and solid bodies. Admittedly, it is difficult to grasp that such a wide variety of phenomena are based on electromagnetic interaction alone: everything from chemical reactions or an electric light turning on to the TV screen, which usually offends our sense of taste; and from the deflection of a compass needle or the reflection of light to the taste of our food. Yet on the small scale of atoms and molecules, electromagnetic forces really do reign supreme.

Unlike gravity—which always attracts—electromagnetic force is contingent on the type of electrical charge. In other words, depending on whether the charges are positive or negative, this force can either be attractive or repulsive. "Likes repel and opposites attract," as we all memorised in our school days, even extending the rule to people's characters. Be that as it may, it is worth underlining that gravitational forces are encountered throughout the vast universe, since they only require the presence of mass to appear. By contrast, no electromagnetic forces are exerted between the heavenly bodies, which are electrically neutral. If the sun and the planets did have some kind of electrical charge, our solar system could scarcely be as stable as it is today. Yet the same mathematical fabric underlies the laws governing the first two forces. Consequently, like gravity, electromagnetic force decreases dramatically as the distance between charges increases.

The third of the forces acting on the universe is not encountered in worlds and phenomena familiar to our experience. To get to know it we have to enter the realm of radioactive decays. The first physicist to cross that perilous threshold was Henri Becquerel, in 1896. Like his father and grandfather, he had worked on the phenomena of fluorescence and phosphorescence. At some point Becquerel noticed that some photographic plates he happened to put away in a drawer had darkened. While others would have ignored the fact, his genius led him to suspect that this was due to some uranium salts lying near the plates, which were emitting a strange kind of radiation. Though he clearly didn't realise it, Becquerel had discovered radiation, i.e. the fact that the nuclei of certain elements are not stable, but spontaneously change into others, emitting nuclear particles and electromagnetic radiation at the same time. Radioactivity was persistently investigated at great personal cost by Marie and Pierre, the mythical Curie couple. It was they who discovered the radioactive elements polonium and radium, and doggedly researched the nature of nuclear radiation. For their work they and Becquerel were jointly awarded the Nobel Prize for Physics in 1903. That same year, having lived in poverty and torment, the talented Marie had only just completed her doctoral dissertation. In 1911, a few years after her husband's death in a street accident, Marie also won the Nobel Prize for Chemistry. She was to go down in history not only as an engaging, ingenious presence, but also as the first to truly rupture the male-dominated world of science.

So radioactive nuclei are exceptional in being mortal, as opposed to the immortality of other stable nuclei. Changes in their composition are usually due to a neutron turning into a proton or vice-versa. This necessitated the existence of another, entirely different force, capable of making nuclear particles change identity. In other words, what was needed was an alchemist, equipped by nature with the appropriate powers.

That alchemist was weak force. But that shouldn't create the impression that its activity is purely maleficent, being linked to nuclear energy and so on. After all, the so-called fusion used by the sun to produce its life-giving energy is also nuclear, as are the processes that lend their everlasting light to the stars.

Weak force appears in nuclear processes and is very short-range, but its role in the universe is in no way inferior to that of the other forces. In fact, the development of large accelerators mainly in the 1970s revealed that weak interaction only likes to have dealings with neutrons and electrons. As for its strength, its name proved correct: weak interaction is a billion times feebler than what we quite rightly call strong force.

This should come as no surprise. Strong interaction is what keeps protons and neutrons tightly bound to atomic nuclei. It has a very short range not exceeding the area of the nucleus, i.e. one centimetre divided by 10^{12} (one followed by twelve zeroes)! Nevertheless, it is to strong interaction that our material world owes its stability. Thanks to the existence of strong force, a table and the book you are holding right now, its author, readers and all the objects surrounding them, the planets and the sun—in short, material bodies in general—do not break down into the protons and neutrons they are composed of. Suffice it to say that this strong force is greater than gravity by a factor of 10^{40}! And since masses in the world of atomic nuclei are extremely small, the gravitational force that does exist is entirely negligible.

So our picture of things now has a descriptive simplicity to it. While the web of interactions extends right across the universe, it is more tightly, differently woven in atomic dimensions. There, three of the interactions intertwine and act via clearly distinguishable, ultra-strong threads. Gravity is left to reign beyond the bounds of the atomic world, at great distances and in the boundless space of the universe. It is thus the driving force behind the evolution of galaxies and stars.

As is only natural, the study of interactions between fundamental particles has predominated in recent decades, becoming the research field for a large number of gifted scientists. Besides, the building of large accelerators has made it possible to conduct close-up experimental studies of the particles active in the microcosm. An unforeseen, wondrous world has emerged from the violent collisions between them, while their Ovid-like metamorphoses have revealed the existence of even more elementary building blocks. Both complex and costly, this research has led to a better understanding of the fundamental forces and expanded the limits of our knowledge.

The distance from what science knew yesterday thus seems far indeed. Unfortunately, the same also applies to the gap between scientifically developed countries and poorer ones. As Pakistani theoretical physicist Abdus Salam recounts: "I still remember the school at Jhang in Pakistan (Jhang is my birthplace). Our teacher spoke of gravitational force. Of course, gravity was well-known, and Newton's name had penetrated even to a place like Jhang. Our teacher then went on to speak of magnetism; he showed us a magnet. Then he said, 'Electricity! Ah, that is a force which does not live in Jhang, it lives only in the capital city of this province, Lahore, 100 miles East.' (And he was right. Electricity came to Jhang *five* years later). And the nuclear force? That was a force which lived only in Europe. It did not live in India (or Pakistan) and we were not to worry about it."[11] Luckily, Salam never stopped worrying about the forces that lived far away from Chang. Many years later, he was to share the Nobel Prize for Physics for unifying the electromagnetic and weak interactions.

That unification was the first grand step on an arduous journey, though the end was still far off. In fact, there is now some suspicion that there may not be an end. What is for certain is that light has always shown the way on that difficult quest.

5.6   On Along the Trail Blazed by Light

The major differences between the four forces acting on the universe should deter physicists from making any attempt to unify them. In other words, from aiming to form a single mathematical model capable of encompassing all the interactions.

At first sight this aim not only appears impossible, but thankless. The main characteristics of the four interactions do not leave much room for believing that they may resemble each other. Their strength varies by numbers in the dozens of zeroes, when moving from strong force to gravity. Besides, their range also shows the same huge differences. Gravity and electromagnetism do not appear to have limits, whereas the weak and strong force are limited to the infinitesimal spaces of the microcosm. Nor do the other characteristics of the interactions point to common origins, either: electromagnetism most definitely requires the presence of charges to manifest itself, while some mass is needed for gravity. As for weak force, neither of the two seems necessary, since neutrons are—in a manner of speaking—particles without charge or appreciable mass. So nature could scarcely be suspected of being simple to describe or parsimonious in the ways it expresses itself.

Yet despite these obvious differences, in the final analysis adherents of a "Theory of Everything" regard the four forces as being manifestations of one single superforce. Though obviously with a fair dose of exaggeration, physicist Paul Davies terms this "the font of all existence." Protons, distant galaxies and even human beings are probably products of this almighty, all-encompassing superforce. The quest is aimed precisely at discovering a unified theory to describe and account for it.

It goes without saying that this quest has also run up against no end of theoretical or even philosophical reservations. Nor have financial parameters been absent, either: following lengthy debate, in 1993 plans were permanently shelved for construction of a super collider in the USA that would have had much to offer essential experimental research, on grounds of cost.

Nevertheless, these difficulties have not proved capable of stopping the age-old quest for some kind of simplicity or unified causation that may lie hidden behind phenomena. Whether the motive for that quest is metaphysical, practical or simply epistemological is of little interest. "God may be subtle, but He is not malicious," said Einstein, stating his conviction that the unified theory did exist. Indeed, he himself spent a large part of his life in a vain quest for It. Lastly, to the end of his life Hawking continued to envision the consequences of its discovery as being idyllic: "Then," he wrote, "we shall all, philosophers, scientists and just ordinary people, be able to take part in the discussion of the question of why it is that we and the universe exist. If we find the answer to that, it would be the ultimate triumph of human reason – for then we would know the mind of God."[12]

While it is doubtful that we will ever know the mind of God, or God himself, it is nevertheless revealing that light gloriously showed the way in that desperate quest. For so it is: after a lengthy period when it only partially applied, quantum electrodynamics became the comprehensive, most accurate description of light's pathways and its relationship to matter particles. So then the logical question was why there shouldn't also be a similar framework for the remaining interactions. Light may not yet have shown us God's mind, but it has at least illuminated one of his facial expressions.

This prospect turned out to be so crucial that the attempt to investigate the other interactions in a manner similar to light has already yielded rich rewards. In fact, in an impressive vindication of their undying faith, theoretical physicists have triumphantly succeeded in fully unifying weak force with electromagnetism.

Nevertheless, we have already seen how the theoretical description of electromagnetic force rested on the bold idea that certain intermediary "carriers" were exchanged between charged particles. These carriers were photons. Known and yet unknown, these electrically uncharged particles took on the role of carrying electromagnetic force. So when investigating the other interactions, the first question posed was whether a similar process of exchange operated or could be devised there, too.

Providing an answer to that calls for an understanding of how matter is composed at its very deepest level, i.e. its elementary building blocks. As is obvious, at such a level one has to look for any possible correlation between photons and forces of exchange. Forces are the medium permitting particles to interact, whereas particles are the units on which forces act. Thus both are facets of the same quantum space, at least as far as the weak and strong interactions are concerned. Having already seen a successful macroscopic interpretation of its behaviour in the general theory of relativity, gravity would perhaps have to be accompanied by other specifications. But they would always lie on the trail blazed by light.

5.7   Heavy Light

In contrast to electromagnetic forces and gravity, the truth is that weak atomic forces are hard to perceive in everyday life. Their presence is limited to radioactive decays and the nuclear processes in stars. Nevertheless, their role in the evolution of the universe is no less significant. Besides, it is thanks to weak interaction that the sun lights and warms the Earth, allowing life to flourish.

One of the most persistent attempts to understand weak force was made by Enrico Fermi, an eminent Italian physicist whose name became associated with the inception of the atomic age. Fermi belonged to that rare breed of scientist possessed of an ingenious theoretical mind and great skill as an experimenter. He was also well versed in the classics, and could recite Dante's *Divine Comedy* and several of Aristotle's works from memory.

In 1938 Fermi was awarded the Nobel Prize in Physics for his papers on radioactivity and nuclear decays, but immediately after the award ceremony in Stockholm he fled to the United States to escape fascism. In his newly adopted homeland he continued his career as a professor at Columbia University. He also led the attempt to build the first nuclear reactor, in an underground squash court beneath an abandoned section of sports field in Chicago. The first controlled nuclear reaction went down in history on 2nd December 1942, marking the inception of the atomic age and rapidly leading to the construction of the first atomic bomb. This age was to permanently scar mankind, unleashing the forces of good and evil and creating ethical and cultural dilemmas. It should be noted that Fermi himself also fell victim to radiation, dying in 1954 of the force he had nevertheless triumphantly tamed.

Of course, none of the above foreshadowed the nature of weak force. The first interpretation of it was given by Fermi's theory, describing the radioactive conversion of a proton in the nucleus into a neutron. Yet despite the fact that the theory offered a satisfactory interpretation of experimental results hitherto, it did not get around certain serious inconsistencies. Today, however, after decades of theoretical and experimental efforts, there is no doubt on one almost inconceivable thing: deep down, the weak force lurking behind radioactive decays is of the same nature as electromagnetism. In the final analysis they constitute a single, unified force. The time may not be far off when school textbooks will follow academic papers in referring to it as the "electroweak" force, just as occurred with the corresponding unification of electricity and magnetism.

This new, daring step in the unification of forces was confirmed by a remarkable experimental discovery concerning the "carrier" of weak interaction, i.e. the particle which, by analogy with quantum electrodynamics, transferred weak force between protons and neutrons, causing them to transform. This carrier not only turned out to exist, but triumphantly emerged from experiment exactly in line with theoretical predictions! It goes without saying that those predictions sprang from the need to formulate a theory possessed of the mathematical and interpretive beauty found in quantum electrodynamics. Ever since light had thrown the road wide open, it wasn't easy for physicists to take the byways.

At first glance, however, it remained uncertain whether that really was the right road. If nothing else, unlike the photon, the carrier of weak force had to have both electric charge and mass; indeed, a mass much greater than that of the proton. Steven Weinberg, who made a decisive contribution to forming the theory, has the following to say: "I worked out a particular concrete realization of this theory, that is, a particular set of equations that governed the way the particles interacted and that would have the Fermi theory as a low-energy approximation. I found in doing this, although it had not been my idea at all to start with, that it turned out to be a theory not only of the weak forces, based on an analogy with electromagnetism; it turned out to be a unified theory of the weak and electromagnetic forces that showed that they were both just different aspects of what subsequently became called an *electroweak* force. The photon, the fundamental particle whose emission and absorption causes electromagnetic forces, was joined in a tight-knit family group with the other photon-like particles predicted by the theory: electrically charged W particles whose exchange produces the weak force of beta radioactivity, and a neutral particle I called the 'Z'…"[13]

The prediction that particles heavier than the proton did exist, and that exchange of them created weak interaction, was confirmed in an epic experiment carried out in 1983 at CERN (the European Organization for Nuclear

Research) in Geneva. Since it called for considerable technical skill and the construction of complex detectors, hundreds of physicists, technicians and collider specialists from different countries joined forces to prepare the experiment and carry it out. Strong collisions between protons and anti-protons (i.e. between matter and antimatter) were thus created in the large underground accelerator, which is 7 km in diameter. These produced the highly prized particles that are the carriers of weak interaction. Note that these particles have an infinitesimally short lifespan, and were tracked down from the products of their decay. In fact, tracking their fleeting presence called for a detector reminiscent of something from the pages of a science fiction novel. Its central unit—a gargantuan system of devices, optical fibres and microprocessors—lay 60 m under the ground, was as tall as a two-storey house and weighed as much as five jumbo jets! But what really matters is that, to the delighted whoops of physicists, the experiment first confirmed the existence of the positive and negatively charged W particle, and later that of the electrically neutral Z particle.

It was thus only fair that the 1984 Nobel Prize for Physics should be awarded to those who had thought up and carried out the grand experiment. And as is ever the case in human history—whether it's the building of the pyramids or a battle fought by generals and rank and file soldiers—out of the thousands of participants, the honours focused on the leaders: Italian experimental physicist Carlo Rubbia, a professor at Harvard University and leading CERN researcher (thanks to the dense transatlantic flight network); and Simon van der Meer, an ingenious accelerator engineer who had developed a system for accumulating antimatter. A few years earlier, not for the first time jeopardizing its reputation, the Nobel Committee had awarded the famous prize to three theoretical physicists who had pioneered the investigation of weak force, and predicted its unification with electromagnetism. They were Steven Weinberg, professor at Harvard University; Abdus Salam, professor of Theoretical Physics at Imperial College London; and Sheldon Glashow, also a professor at Harvard.

Nevertheless, reasonable doubts may well have arisen in some readers' minds. From what we have actually said it can be concluded that W and Z particles are in some sense brothers to the photon. At first glance they have nothing in common: the photon has a mass of zero, while theirs is extremely large in microcosmic terms—roughly 90 times that of a proton. What is more, they are electrically charged, whereas photons are neutral. So observers would be unlikely to attribute a common nature and mechanism to two forces with such different carriers.

Physicists, however, are no ordinary observers. In fact, their inherent need for a unified schema brushed aside every difference emerging from experimental data. So although it has been proved that the electromagnetic and weak forces can indeed be described by a unified mathematical theory, in present-day conditions in our world the theory displays an attribute called "spontaneous symmetry breaking". This fine-sounding expression, which translates into elegant and complex mathematics, means something quite simple: when observed at low energies, the photon and the carriers of weak force appear to be different types of particles. Nevertheless, under suitable conditions they are seen to be one and the same: at very high energy levels and temperatures, the new particles acquire zero mass and behave like photons. So although the three large mass particles—positive and negative W and neutral Z—appear in different guises, they are once again light. Just that this time they are heavy light!

It is obvious that heavy light simplifies things as regards the peculiarities of weak force. Yet one crucial question remains: is its unification with electromagnetism no more than a wonderful theoretical tool, or did such a daring concept also "work" at some point in reality? In other words, can the colossal temperatures and energy levels needed to manifest unified force be achieved in our world, or are we talking about a more or less virtual schema, which only exists in our minds?

The answer is categorical. At least at present technological capabilities, creating conditions of high enough energies and temperatures to lead the forces to act in unison is unachievable. That would require an enormous accelerator, reaching as far as the nearest stars. But there was one moment in the history of the universe when suitable conditions did apply—and that was none other than he moment the universe was created!

As science now agrees, the universe came into being through an enormous explosion, unimaginable in terms of its feel or force. According to existing indications, the renowned Big Bang took place around 13.8 billion years ago. In the first few milliseconds after the Bang, temperatures were exceptionally high and matter particles were seething with activity. It is only in these conditions that the now lost unity of nature can come into being. At that point, electromagnetism and weak force constituted one single force. In other words, they were two wondrous gifts from the Maker rolled into one. As the universe began to expand from minute dimensions and cool down, the two forces separated, each taking on its own role from then on.

The sensational thing is that several billion years after the Big Bang, a number of gifted individuals living on an insignificant planet were engaged in a persistent quest for the unified theory of the cosmos. In doing so, they discovered the common origin and nature of weak and electromagnetic force—two forces which now appear entirely different. The road to that discovery had been shown by light and the way it interacted with matter.

5.8   Quarks and Their Eternal Bondage

At least as regards its philosophical content, the now dominant view of how matter is structured does not greatly differ from that conceived by Democritus. The only thing is that atoms are no longer regarded as the building blocks of the material world, as they were for many centuries. The same applies to protons and neutrons, which were later revealed to constitute atomic nuclei. At an even deeper level, matter is made up of fundamental, indivisible constituents known as quarks. Their strange name originates in the James Joyce novel *Finnegan's Wake*, perhaps to show—in vain—that physicists are not entirely ignorant of literature. Quarks, then, are the minutest constituents of the material world, and they come in three kinds. In fact, as sacred law decrees, they also have their own corresponding antiparticles, the antiquarks.

Just like atoms in the past, each kind of quark obviously has different properties, on the basis of which it is assigned a "flavour". Of course, this has nothing in common with flavours of ice cream or sorbet, it is a natural trait that determines permissible combinations. Three quarks—two of which are the same flavour—make up the protons we are familiar with. A different combination, yet again of three quarks, makes up the neutron.

All the fundamental particles detected over decades in cosmic rays or artificially generated in accelerators turned out to be complex. In other words, they were combinations of two or three flavours of quark. In fact, they were named *hadrons*, after the Ancient Greek root *hadr-os* [="thick, heavy"]. In the other camp were the so-called *leptons*, namely the electron and the neutron; though highly abundant in the universe, the neutron had no part to play in the structure of ordinary matter. Thus, a minimal number of quarks and leptons, and just four fundamental interactions, were all that sufficed to understand our world. Its apparent complexity once again hid in its depths a simple, symmetrical structure.

Nevertheless, the problems that remained were far from insignificant. The main one lay in the fact that while all the experiments indirectly confirmed the existence of free quarks in the interior of hadrons, none had been found! Physicists looked for them everywhere, from lunar rocks to cosmic rays—all in vain. Indeed, when stationary protons were bombarded with other particles that had gained enormous speeds in accelerators, the fragments were not, as anticipated, any of the quarks. They were once again complex matter particles. It was if someone had taken aim at an opaque bottle, though one which they knew to contain three small balls. To their surprise, hitting it only led to the creation of new bottles, which yet again contained some of the balls!

As we have already mentioned, the force holding quarks in the interior of a proton or neutron is exceptionally strong. All the same, it does have one major quirk: unlike electromagnetism or gravity, it increases indefinitely with distance! So quarks move relatively freely in the interior of hadrons, but if we were to try and release them, the force of attraction between them would be impossible to overcome. They are like prison inmates who feel relatively free in their cells, but who find high walls and electric fences standing in their way if they attempt to escape. Certain quantum field theories provide an interpretation for this unexpected behaviour of strong force, the intensity of which tends to infinity as the distance between quarks increases; it is aptly named "asymptotic freedom". Thanks to it, quarks are unable to separate, and remain permanently imprisoned in the interior of hadrons. "Seen from another angle," as the eminent physicist and professor at the École Normale Supérieure John Iliopoulos once put it to me, "asymptotic freedom becomes true slavery. The price of freedom is infinite."[14]

Modern science had thus reached a conclusion that would excite certain philosophers: the existence of quarks was almost certain, but science's inability to observe them as autonomous entities was equally certain. A similar phenomenon was known to classical physics: magnetic monopoles do not exist. A magnet cut in two results in two entire magnets, not two separate poles. At all events, on another level the asymptotic freedom of quarks recalled the relationship between the author and the woman who kept accompanying him in silence and then leaving once more. The attraction he felt for her increased the further apart they were, as her face lost itself in constant mirages of light. The author now knew that final separation held no meaning, and that in any case it could never come about.

As is obvious, together with the paradoxical behaviour of quarks (which remain eternally imprisoned in nuclear matter), describing the interactions between them constituted a true challenge for theoretical physics. Researchers already had an important instrument at their disposal in the form of quantum electrodynamics, which yielded elegant and highly accurate descriptions of the interactions between light and matter. The theory's famed diagrams and their mathematical formalism expressed the basic idea that light "quanta" were constantly being exchanged between charged matter particles such as electrons. It was this constant emission and absorption of photons that created the electromagnetic force dominating the world of atoms and molecules. Weak force was also interpreted in a similar way, though in that case heavy light was being exchanged! It thus tied its fortune once and for all to electromagnetism: those whom God has joined together, let no man set asunder!

Consequently, it was only reasonable that when investigating the behaviour of quarks and the strong force that held them in the proton or the neutron, models similar to those in quantum electrodynamics were sought out. So by analogy with photons, physicists conjectured that the world of quarks communicated and was held together by some special messenger particles. Precisely because they stick quarks to each other, these particles were named gluons, from the Ancient Greek *gloios* (="sticky, gluey"). The gluons are eight in number, and like photons must have zero mass. So while photons are responsible for creating electromagnetic force, in a similar way gluon exchange between quarks is the root cause of the strong force dominating the world of the nucleus. Once again, we are on along the trail blazed by light!

All the same, unlike gravity or the electromagnetic force, nuclear force increases in relation to distance. Describing this unexpected behaviour called for additional parameters, as its asymptotic freedom necessitated. So in addition to flavour, physicists hypothesized that each type of quark was characterised by one further interesting trait: its colour. Of course, quark colours do not resemble those of everyday life. They simply play a role similar to the positive or negative charges borne by electromagnetic forces.

Before readers begin to doubt the seriousness of modern science, it is important to stress the obvious fact that beneath these simple, vivid ideas lurk complex, elegant mathematics and fundamental concepts in physics, such as that of fields. In fact, the theory developed to describe the behaviour of hadrons goes by the weighty name of Quantum Chromodynamics, exactly recalling the analogy with quantum electrodynamics, and the fact that in this case colour plays the role of electric charge. Though not without its weaknesses, quantum chromodynamics now stands as the undisputed theory of strong interactions. Deep down, it is once again a choreography of light, though this time in a feast of colours.

5.9   An Unfinished Theory
of Everything—And Its Triumph

We have seen how a theory uniting the weak and electromagnetic forces already exists and has been admirably confirmed by experiment. The quantum chromodynamics describing strong force was also created within the same context, so the effort to combine the two was a matter of course. Thus arose the "Grand Unified Theories" that dominated the landscape of physics for so long. Though their name does connote the obsession with a unified theoretical schema, it is rather misleading; for all their significant successes, the theories have not proved capable of including gravity in their unified picture of the world.

Yet this last force is vital if a Theory of Everything is ever to be formulated. Optimistically speaking, such a theory will encompass all the fundamental processes in the world in one single mathematical equation or—if needs be— in just a few. The solution to those equations will, without doubt, describe our universe. In other words, it will have the three dimensions of space and the fourth temporal one; it will include quarks, electrons and photons, making up the stars, the planets and our own selves. Through their action, gravity and the electromagnetic or atomic forces will compose matter in its various forms. A Theory of Everything like this will even include the Big Bang from which everything originated, while revealing the internal link between quantum mechanics and gravity. "Concepts of physics as we know them today will be completely changed as the story unfolds,"[15] claims Edward Witten, one of those most optimistic that a Theory of Everything will be found. Having contributed so much to the quest, Stephen Hawking predicted some years ago that a Theory would be formulated within the next couple of decades. "I repeat the same prediction," he quipped during a lecture at the University of Crete in 1998, "only the couple of decades start from now!"

So while the quest for—or the dream of—a Grand Unified Theory intensified over the closing decades of the twentieth century, the results were not promising, and scientists' enthusiasm had begun to wane. Firstly, because gravity was stubbornly resisting incorporation into a unified schema, and secondly because the theories formulated were plagued by multiple arbitrary assumptions and the impossibility of testing by experiment.

All the same, the efforts made by a multitude of newcomer and distinguished researchers to formulate a unified theory did not go to waste. Far from it. Out of the chaos of fundamental particles and the attempt at a mathematical description of them arose an ingenious and consistent theory, which was to become known as the Standard Model of interactions. While gravity still did not fit into this unified schema, the three other forces—electromagnetism and the strong and weak force—squared neatly with its requirements. The Standard Model even demonstrated its virtues, since among other things it predicted the existence of an additional quark, and precisely calculated the masses of the W and Z bearers of weak interaction, which were triumphantly confirmed by experiment.

It does of course go without saying that the mathematical formalism of the Standard Model is anything but simple. It rests on the concept of the "quantum field" expressed and created by the mediators of interactions: photons, gluons, positive or negative W and neutral Z.

Simply recording these mediators highlights an imbalance: photons and gluons are without mass, whereas the remaining ones not only have mass, but are far heavier than the protons familiar to us all. But this is where a phenomenon far from uncommon in the history of physics comes to the fore. It involves the formulation of a daring prediction, and the need that creates for experimental verification.

So it was. From as early as 1964, while studying spontaneous symmetry breaking, the British theoretical physicist Peter Higgs predicted the existence of a particle and corresponding quantum field with an important mission: to enable the particles making up the Standard Model to gain their much-coveted mass. This new particle, belonging to the category of bosons—i.e. those with zero or integer spin—was to go down in physics scholarship as the Higgs boson. But thanks to the ingenuity of a publisher who had brought out a book on the topic, entitled *The God Particle*, that was the name by which the astounded public came to know it after its experimental verification. From the title, many readers might suppose that the God that humans have been vainly searching for down the centuries had finally been found.

But just like the search for God, experimental observation of the Higgs boson was no trivial matter. Producing it meant bringing about a proton collision at enormous energy levels, in the presence of sensitive detectors to record and process "suspect" events at the collision points. It was eventually possible to build the complex apparatus, at untold cost and with a unique combination of human abilities, at the European Atomic Research Centre-the world-famous CERN in Geneva. I should perhaps say that I had the enormous luck and pleasure of working there for an entire year as a visiting researcher, and will never forget my experiences and sense of wonder at the centre's unique environment and research achievements.

The CERN Large Hadron Collider—also known by its initials as LHC—which shouldered the onerous responsibility of detecting the existence of the

Higgs boson, surpasses the wildest human imagination in terms of construction and operation. The main part consists of a circular tunnel with a circumference of 27 km, at a depth of 150 m below the France-Switzerland border. Inside the tunnel, two beams of protons travel in opposite directions, gaining incredible energy levels—comparable to those prevailing shortly after the Big Bang—before colliding head on. The barrage of subatomic particles produced at the collision points is recorded and analysed by complex sensors and high-powered computers. The scale of the venture can also be seen in some characteristic numbers: construction of the Large Collider took an entire decade, from 1998 to 2008, in an undertaking that required the collaboration of around 12,000 physicists belonging to universities and research institutions in at least 100 countries. All mention of the cost is avoided, since it would feed the age-old divide among readers between those who believe the enormous sum would have been better used for social ends, and others who believe that the value of scientific progress is unique and non-negotiable.

Yet above and beyond that, one fact remains. That in 2012—in other words, almost 50 years after the Higgs boson was predicted—its existence was confirmed by the Large Hadron Collider and the sensors attached at the points where the protons violently collided. In fact, a preliminary estimate of its characteristics was also gained: its mass was over 130 times greater than that of the proton, but it only had an infinitesimal lifespan before breaking down into other, lighter particles.

The discovery of the Higgs particle was a triumph for the Standard Model, filling the scientific world with enthusiasm. Indeed, as was anticipated, in 2013 the Nobel Prize for Physics went to its originator, who was already well into his 84th year. Peter Higgs shared the prize with another leading physicist, Belgian François Englert, who had also contributed to highlighting the mass creation mechanism via spontaneous symmetry breaking.

All the same, one defining truth should not be forgotten: the beginning of the road leading to the Standard Model and its successive goes back to electromagnetic interaction and the explanation for it, i.e. that charged particles exchange photons. In other words, as in other things of the world and life, it was light that showed the way.

5.10   Two Parallels That Converge

Of course, it goes without saying that discovery of the Higgs boson raised the prestige and acceptance of the Standard Model to new heights. In fact, it's worth noting that in addition to its experimental successes, the Model made a significant contribution to understanding the structure and evolution of the universe. That was because it predicted the convergence of interactions at high energy levels, thus offering a cogent picture of the first moments of creation.

A fundamental precondition for that picture was that the moment the Big Bang created the universe, the three forces making up the Standard Model were unified with gravity into one single superforce. As the universe began to expand and cool, gravity separated from the other forces first and began to rule the macrocosm. The other three forces that were to dominate the microcosm also gradually split up. On the other hand, the ubiquitous and powerful Higgs field enabled atomic particles and the bearers of interactions to acquire mass and slowly take on their special mission. The first nuclei of matter only took a few minutes to form, whereas hundreds of thousands of years had to pass for electrons to bind to them so that the first neutral atoms could exist.

So one obvious conclusion arises from modern cosmology, which may be one of the most significant leaps in human thought: the microcosm of matter and the macrocosm of the universe constitute two parallels. Nevertheless, they are two parallels that converge.

Yet for all its triumphs, the Standard Model is not without serious weaknesses. It is, for instance, incapable of interpreting the existence of dark matter, which according to the latest data dominated in the creation of the universe. Nor does the model offer any convincing interpretation of the neutrino oscillations that have puzzled physicists. Its main weakness, however, lies in the refusal of gravity to fit into the unified schema. So while its wondrous structure and predictive ability are not called into question, the Standard Model is far from satisfying our dream of a Unified Theory of our world. At best, it is an "unfinished" symphony, which may be completed in the future. Yet like Beethoven's Tenth Symphony—which is likewise unfinished—the Standard Model does not lack beauty or expressive power. It's worth noting that the Tenth Symphony was recently "finished" with the aid of Artificial Intelligence. And those who have listened to the amazing "supplement" have to say that it is on a par with its "unfinished" model in evocative power or rhythm. But is that also a token of optimism for a "Unified" Theory that remains unfinished nonetheless, as it fails to include gravity? That is not easy to assess. All the same, I don't believe that Artificial Intelligence will be the one to achieve much-coveted unification. We are more in need of a new Einstein, who will view the problem with his or her own discerning gaze and thought.

It is of course obvious that scientists have also tried other, different paths to arrive at the unification of gravity with the other forces in the Cosmos. Thus, for a long time so called String Theory was popular, resting on the revolutionary hypothesis that the basic elements in our cosmos were not elementary matter particles, such as quarks and electrons. Instead, they were paradoxical entities which only had length and which, if observable, would resemble infinitely thin strings. Like the strings of a violin, the ones in the theory vibrated at differing rates in space–time; but rather than producing music they produced energy and, consequently, matter. So that was where the elementary matter particles making up the entire universe originated from.

At any rate, while String Theory appeared to meet many of the criteria for a Unified Theory, also incorporating gravity in its framework, it rapidly ran up against intractable difficulties. Thus, to fully reveal its indisputable beauty, String Theory called for a ten- or even twenty-six-dimensional universe! Its main weakness, however, was that experimental verification of it was impossible. As the hallowed rule of natural science demands, no theory can be accepted if not confirmed by experiment. It is the rule that sets science apart from ideological or philosophical trends, and underscores its potential grandeur.

6

Light Shakes Hands with Life and Art

6.1   Light, the Driving Force of Life

Light blazed the trail for all the major discoveries in physics, leading mankind on a step-by-step journey to meet the paradoxical quantum world. Yet it is also vital for one other reason; in essence, light is the driving force of life.

As might be expected, this formidable capacity rests on solar radiation, the main source of light's presence in our everyday life. And it is a capacity manifested on many levels. Solar energy provides a temperature suitable for the survival of life on Earth, as well as climatic conditions favourable to that end. What is more, part of that same energy has been transformed into deposits of crude oil, coal and various natural gases, which are the workhorses powering modern societies. Only about one billionth of the colossal energy generated by the sun ever reaches the Earth, yet that tiny proportion is enough to sustain and promote activities ranging from basic life functions to the flight of a rocket, and from the light of day to the evaporation of water from the seas. So the famous Greek author Nikos Kazantzakis quite rightly opens his own ambitious modern sequel to the *Odyssey* with the wonderful lines:

O Sun, great Oriental, my proud mind's golden cap,
  I love to wear you cocked askew, to play and burst
  in song throughout our lives, and so rejoice our hearts[1]

The primary process maintaining life on Earth is, of course, photosynthesis. As its etymology suggests, this word means synthesis via light—an altogether amazing manifestation of the hidden processes required for the complex phenomenon of life to flourish.

The final word in photosynthesis is had by plants and their wondrous functions. Only a tiny fraction of solar energy, not exceeding 0.05% of the total, takes part in photosynthesis. Yet this is sufficient to serve as a cue for the perpetual role played out by plants. In the modern world it is a role we do not always appreciate. In the pursuit of short-term profit, humans often turn against the plant kingdom. Yet if plants were ever to go on indefinite strike, in a fit of pique over our thoughtless stance, the human species and all its wrongdoings would fast disappear.

At least as regards light, one characteristic thing about plants is that some of their behavioural traits mimic those of humans. For example, if they are indoors they will turn towards the window to catch the sun's rays. Outside too, growing unimpeded, they grow or orient their leaves in search of the best way of enjoying sunlight. All plants are in any case enigmatic biological clocks, with a complex mechanism regulated by light. Whether this light comes from the sun or from an electric bulb is of little importance; plants will move their leaves and produce blossom simply because they interpret the physical properties of light, its energy or its wavelength.

It follows that the life of plants—and, indirectly, the life of every living organism—is in essence dependent on light. To provide for this vital function, nature has equipped plants with chlorophyll, an invaluable absorbent substance that enables them to trap light. Chlorophyll acts as both a receiver and an analyser, by reflecting green light rays while absorbing red ones. In fact, it is to this property that plants owe their typical green coloration. But the important thing is that the light energy absorbed by chlorophyll triggers chemical reactions, resulting in the synthesis of organic compounds and oxygen.

So photosynthesis has a key mission. It produces life-supporting oxygen, as well as the carbohydrates needed by plants and animals: starch, sugars and cellulite. Yet while carbohydrates are photosynthesised by plants, those very same plants consume part of them in order to grow. For that matter, all living beings burn carbohydrates. Photosynthesis is what leads the cyclic flow of matter—carbon, oxygen and, indirectly, other elements—through the biosphere. Carbohydrates have even wormed their way into our everyday vocabulary, since nowadays people are constantly bombarded with dietary advice, most of which is utter nonsense.

All humans, plants and animals thus participate in the carbon dioxide cycle, which, together with water, is the raw material for photosynthesis. If we take into account that carbon dioxide is also produced by all the various engines that burn diesel or petrol, then the amount of that gas converted into carbohydrates via photosynthesis is mind-boggling—around 200 billion tons per year! This colossal energy undertaking, which photosynthesizes in a perpetual cycle, uses a leaf area of about 75 million square kilometres. Looked at this way, the claims made by photosynthesis pioneer Eugene Rabinowitch do not sound all that overblown. "Physiologically speaking," he writes, "all the animals on land and in the sea, including man, are but a small brood of parasites living off the great body of the plant kingdom. If plants could express themselves, they would probably have the same low opinion of animals as we have of fleas and tapeworms."[2]

Not content with being endemic parasites on the plant kingdom, humans are dangerous parasites. Our aggression is not only directed against fellow humans, but also against the entire ecosystem on our treasured planet. The plant world has not escaped our destructive mania: large cities often suffer from a lack of greenery, which offers respite and adds to the quality of everyday life. Even greater is the destruction wrought on the world's forest riches. For instance, tropical forests are being ravaged at terrifying speed in the name of so-called development needs. Yet while they cover a very small part of the Earth's surface, these forests play host to more than half of all the species in the animal kingdom, and absorb part of the carbon dioxide being so thoughtlessly pumped into the atmosphere by human activity, cars and industry. If this seemingly unstoppable siege of the plant kingdom continues, the overabundance of carbon dioxide in the atmosphere will lead to runaway global warming, which is most probably already in the offing. Our delicate thermal equilibrium will then be upset, with dire consequences for the planet. The freak weather conditions already being seen in country after country may well be a dramatic warning.

Light, then, works in harmony with the plant world to maintain the perpetual cycle of photosynthesis that is vital for life. Yet while in evolutionary terms humans stand at the pinnacle of life, they have thoughtlessly turned against the environment that hosts both them and life in all its variety. Though difficult to interpret, this paradox is beyond doubt the tragedy of our civilisation.

Numerous reservations have been expressed in the past, but an enormous climate crisis is now a painful reality for our planet. Humankind must accept its responsibility and do what is needed to tackle it in what time remains—if there is any!

6.2   The Incredible Miracle of Sight

It is via light that humans reconstruct the world around them, investigating its forms and movements. This important function is performed by sight, which, as we all learnt at school, is one of our basic senses.

Living beings acquired and perfected sight over the lengthy course of evolution. But since it is always based on electromagnetic waves reaching and being detected by the relevant sense organs in each organism, it goes without saying that vision varies widely in terms of both structure and capabilities. Its culmination in mankind stands as a complex and wondrous mechanism. As Leonardo da Vinci writes: "O excellent thing, superior to all others created by God! What praises can do justice to your nobility? What peoples, what tongues will fully describe your function? [...] Owing to the eye the soul is content to stay in its bodily prison, for without it such bodily prison is torture. [...] Who would believe that so small a space could contain the images of all the universe?"[3] It is worth noting that as a pioneer in both art and science, da Vinci himself looked on the world with incredible insight and inspiration.

So how do we see? The numerous answers to that key question recorded down the course of the centuries have been no more than approaches to the truth. Mainly after the scientific revolution, the structure of the eye and the process of vision became a serious object of study among scientists, doctors and philosophers. Together with the growth of microelectronics, the progress seen in neuroscience over recent decades has both literally and metaphorically shed light on many points in the complex function of sight. Let's just note that human sight is called upon to perform and combine the distinct, demanding functions of seeing shapes, judging distance and movement and, lastly, perceiving colour.

The basic mechanism is superficially describable. Thus, for us to see a bird in flight the lens in our eye forms an inverted image of it on a natural screen called the retina. This has millions of light-sensitive cells, which break the image down into its different constituents. The bird's motion, shape and colours are initially apprehended as photons and codified into weak electric signals, which head via optic nerves to the cerebral cortex. This process requires countless calculations, far beyond the capabilities of a computer. The brain then immediately translates the signals to create a perception of the bird—standing upright, with its rich colouring and characteristic details.

Via sight, then, the brain synthesises visual stimuli into images of the beauty, mystery or ugliness in our world. Yet science can only partially grasp this complex process. The British philosopher and mathematician Alfred

North Whitehead offers an uncompromising summary of the role played by the brain: "… there is no light or colour as a fact in external nature. There is merely motion of material. Again, when the light enters your eyes and falls on the retina, there is merely motion of material. Then your nerves are affected and your brain is affected, and again this is merely motion of material. [...] Thus the bodies are perceived as with qualities which in reality do not belong to them, qualities which in fact are purely the offspring of the mind. Thus nature gets credit which should in truth be reserved for ourselves: the rose for its scent: the nightingale for his song: and the sun for his radiance. The poets are entirely mistaken. They should address their lyrics to themselves…"[4]

It is often said that the basic mechanism of sight bears many similarities to a camera, or better still to a modern camcorder. In reality, however, vision emerges as far superior. The human eye does indeed resemble a camcorder, yet it is immeasurably more sophisticated. After all, it has taken billions of years of experiments and trials to construct. Weighing just 7 grammes, it is only a few cubic centimetres in volume and, what's more, has no need of electricity or batteries to operate. Yet it has an extremely wide field of view, and does not distort images in the slightest; it can revolve at will and follow moving objects without difficulty; record distances unaided by ultrasonics and render colours perfectly, even in low-light conditions; and lastly, it boasts the fastest and most accurate autofocus capability. So no camcorder in the near or distant future will be able to compete with the strengths of the human eye. Anatomically speaking, it should be noted that the eyes are an extension of the brain, which is why they are located at one of the highest points in the body. In essence, though, they are a window to the soul and offer more vital information than any other sense. So poets such as Octavio Paz are perfectly justified in talking about the eyes of their beloved:

> Your eyes are the country of lightning and of tears,
>   of silence that speaks,
>   of storms without wind and the sea without waves,
>   caged birds and dreaming brutes,
>   [...]
>   a basket of the fruits of fire
>   and a nourishing lie,
>   the mirrors of this world and the doors of more to come…[5]

At any rate, the incredible mechanism summarily known as sight has a number of recognisable constituents. The lens in the eye, a marvel of micromechanics, is responsible for focusing. It is held in place by a ring-shaped muscle of innumerable fine strands thinner than a hair. Object images are produced by photons that excite the light-sensitive cells in the retina. These so-called photoreceptors cover the back of the retina like a mosaic, and are divided into two categories known as cones and rods. Though named after their shape, this does point to an essential difference: cones are responsible for colour vision, whereas rods can only discern shades of grey. So it is as if two films are running in tandem on the retina. The black and white one has the rods as its receptors, while the colour one has the cones.

There are a "mere" six million cones in the eye, concentrated around the centre of the retina. These are actually divided into three types, each of which is selectively sensitive to just one colour, be it red, green or blue. Combinations of these colours generate the sense of coloured vision, just as occurs in a colour TV image. In fact, the genes responsible for colour have recently been decoded. Any defect or hereditary deficiency in these genes leads to colour blindness, in which the world looks black and white.

It's worth noting that simple observation alone does not always reveal the sumptuous world of colours. Most stars appear white, since the light they emit is not sufficient to excite the cones. However, the true colour of stars can be seen in time-exposure photographs of them: the coldest stars are red, while warmer ones are blue and purple. It is also a fact that the cones in women's eyes are more easily excited than those in men, so women see the stars in the night sky in fuller colour than men. This is yet another instance of superiority in women, a further fragment of their particular psyche.

At any rate, our sight is most sensitive in the grey-black zone. There the lead role is played by the rods, numbering over 100 million in each eye! They are located around the outer edge of the retina and are, in a manner of speaking, motion receptors. In moonlight, where only rods are involved in vision, the world has a colourless, dark look to it. Note that the protein molecules comprising rods are kinked in the middle, but when stimulated by light they unravel into a straight line in space. This change in shape triggers a system of electrical pulses and neurotransmitters, which sends a high-speed visual message to the relevant sections in the brain. It should be stressed, however, that the rod mechanism is thousands of times better at distinguishing light than the best modern camera: in some cases the human eye is capable of distinguishing a single photon.

The important thing is that this delicate process initiated by the lens and then the retina is a mere introduction or prelude to the symphony of sight. The symphony itself is the work of the brain: the cones and rods are linked to the relevant sections in it via a complex optic nerve resembling a kind of million-strand telecommunications cable. After a few intermediate steps, the optical information reaches the visual cortex in the form of weak electric currents. The cortex is a vital area of the brain roughly the same dimensions as a credit card, and it is where the electrical signals are processed and the final image put together. Reaching this endpoint is obviously an exceptionally complex process. Suffice it to say that certain areas of the visual cortex act as a phone switchboard, allocating messages accordingly: motion is recognised in some areas, others process colour, while still others analyse image depth or the component parts of shape. Lastly, identifying an object and locating it in space calls for the collaboration of various brain areas. At any rate, it appears that recomposing the electrical signal into an image and recognising whether it is a head or a branch, for instance, occur simultaneously. Perceiving the sum total, i.e. whether it is a person or a tree, is likewise simultaneous.

It is worth underlining once again that as a whole, sight amounts to more than the sum of the individual processes, which are not themselves always comprehensible. It does indeed resemble a wonderful symphony, it's just that the score to it has many indistinguishable phrases, and the aesthetic outcome is almost impossible to account for or express in simple terms.

What should also be stressed is that the stunning precision typical of human vision—as well as that of other living beings—has one serious limitation: it can only detect a thin sliver in the broad spectrum of light radiation. Indeed, the area of the spectrum visible to humans is limited to wavelengths of between 0.4 and 0.8 μm (millionths of a metre) or 4,000 and 8,000 Å. On the other hand, the entire electromagnetic spectrum begins with gamma rays, with wavelengths in the billionths of a metre, and ends with radio and television waves several thousands of metres long.

So our window on the world is but a narrow slit. Yet from that narrow slit humans see the people and things in life, rejoice in the colours of flowers or a painting, and even marvel at the starry sky in all its infinity. Through that sliver of light, she too once appeared on the author's horizon. Whether it was the real horizon or just his expectations, from then on she has accompanied him in silence, often letting her hands play with the light and its shadows.

6.3   And Man Said: "Let There Be Light!"

In the long course charted by humans, the catalytic presence of light has been a source of myths and wonderment. Yet alongside this, the creation of artificial light, banishing darkness and night, has changed our way of life and daily routine.

Fire was the first artificial light source. Though its discovery lies in the distant palaeolithic past, reliable techniques for exploiting it only appear to have been mastered by neolithic humans. Since then, the uses of fire have both literally and metaphorically illuminated the path for societies and culture, bolstering technological progress. So the pride and bitterness expressed by Aeschylus' Prometheus are justifiable when, though bound in chains, he enumerates his acts of beneficence to humans:

> PROMETHEUS: I gave them fire.
>   CHORUS: What? Men, whose life is but a day,
>   Possess already the hot radiance of fire?
>   PROMETHEUS: They do; and with it they shall master many crafts.[6]

Even early on, burning wood was not the only way of producing fire and its gifts. Torches burning rags or tar were used in classical antiquity, and the Romans later lit their towns in the same way. Burning animal fats or oil is another extremely old lighting technique, which led to lamp making. One of the most ancient lamps ever discovered is over 15,000 years old! When solidified, oil also comes in the form of wax, which has a long history and is still used today—whether in emergencies or romantic moments. Candlelight was used by C. P. Cavafy in an agonizing symbolisation of passing time:

> Days gone by fall behind us,
>   a gloomy line of snuffed-out candles;
>   the nearest are smoking still,
>   cold, melted and bent.[7]

Poetry aside, one characteristic fact is that in the Middle Ages candles were so expensive that during trials people had to pay judges a special tax to use them.

With the right techniques, the heat of sunlight can also be used to light fire. The Olympic Flame was lit with the aid of a mirror, while conveying it by torches had a profound symbolism to it; at the 2004 Athens Olympic Games, the Greeks were able to experience both the evocative and the droll moments in the modern version of the ritual. At any rate, Archimedes was the one to use sunlight in a manner as impressive as it was unorthodox. There appears to be a historical basis to his having used a series of mirrors to set fire to a Roman fleet threatening his native Syracuse.

One milestone for artificial light was the Pharos of Alexandria, the oldest lighthouse in recorded history. An apparently elaborate monumental structure 80 m high, its light came from burning wood, and it was equipped with all the scientific innovations and achievements of the time. "Never, in the history of architecture," notes E. M. Forster, "has a secular building been thus worshipped and taken on a spiritual life of its own. It beaconed to the imagination, not only to ships at sea, and long after its light was extinguished, memories of it glowed in the minds of men."[8]

In the nineteenth century it became common for cities and towns to be lit using gas, produced by carbonizing coal. Gas lamps initially produced light via a simple burning jet, but their brightness was greatly increased by fitting a mantle impregnated with thorium nitrate. This burned white hot to emit a characteristic bright light, illuminating people and streets.

It goes without saying that the proliferation of scientific knowledge opened up ever more capabilities for light sources and their uses. The major, most substantial turning point came with the discovery of electricity and the impressive spread of its applications. Nowadays, with the infinite functional and aesthetic variety of electric lighting methods, it is difficult to grasp that the light bulb was invented little over a century ago. On that historic day in 1879, cheered on by an amazed audience, the ingenious Thomas Edison demonstrated the first electric lamp. As a contemporary newspaper report characteristically put it: "Edison's laboratory was tonight thrown open to the general public for the inspection of his electric light. Extra trains were run from east and west, and notwithstanding the stormy weather, hundreds of persons availed themselves of the privilege. The laboratory was brilliantly illuminated with twenty-five electric lamps, the office and counting room with eight, and twenty others were distributed in the street leading to the depot and in some of the adjoining houses. The entire system was explained in detail by Edison and his assistants, and the light was subjected to a variety of tests."[9] Edison's lamp was soon in widespread use, but for some time hotels had to remind their clients that their room light lit up on its own, without matches! Here we should note that Edison had also invented the phonograph and the telegraph, and made significant improvements to the first telephone.

Bulbs and other sources of electric lighting were fast in developing. Light was initially generated by using an incandescent filament. The filament's role was subsequently taken over by low pressure gas that either gives off light or makes a phosphorus coating glow. But a substantial breakthrough in bulb technology was achieved with LED (light-emitting diode) lamps, which are far cheaper to run. Based on the use of semiconductors, when current passes through they generate light in the ultraviolet, visible or infrared part of the spectrum, depending on their chemical composition.

So much for technical details. The fact is that the spread of the various lighting means has not only changed human everyday life, but also its prospects. At every step in our lives we encounter artificial light, varying widely in intensity, colour or the ease with which it serves our needs. Workplaces and shops have their own lighting schemes; roads are passable at night; passengers can even read on the metro; and nightlife "entertainment"—if that's what it is—makes use of beams and spotlights.

Besides, we have no need of further witnesses. The importance of light in our everyday life was recently thrown into relief by the war in Ukraine and the resultant need to curb energy consumption across Europe. Limitations on public lighting and financial pressure to reduce domestic needs showed everyone the extent to which modern life depends on light energy and its multiple applications. The abundance of artificial life banishes the dark, but also has serious repercussions: it distances people even further from nature and its whisperings. Plentiful natural light is considered an advantage with a hefty price tag in modern homes, while even seeing the night sky is now impossible in big cities. Whether it is the gentleness of dusk or the brightness of a summer day, one thing characteristic of modern life is a tormenting nostalgia for natural light. But alongside that nostalgia for light, the author often been tormented by nostalgia for her. At times uncertain whether he had ever met her, he was afraid she might simply have been a trick of the light, that her face and movements were due to light and its reflections.

6.4   Humans Tame Light

From early on, humans felt the need to tame light, searching for ways and means of using it to improve their everyday life and social development. This has led to the invention of various optical instruments that exploit the properties of light and form images of objects to serve people's needs, or sometimes their coquetry! Mirrors have a long prehistory to them. Originally bronze, they began to be made of pure glass from the fourteenth century onwards. Like the present-day versions they had a metallic coating on the back surface to reflect light, and were soon followed by convex and concave mirrors. The ever-present refraction of light is also exploited by lenses. Whether diverging or converging, they focus light beams and magnify objects or bring them closer up to the eye. Lenses have an extremely wide range of applications, and various kinds of them are encountered in glasses and contact lenses, binoculars and cameras. Making lenses has never been a simple matter, since getting them to focus well has always been vital; light rays are often forced to take unexpected pathways. But advances in computers have made complex calculations and difficult approaches possible, lending new impetus to lens designs and combinations of them.

Among all optical instruments, the telescope and microscope are of particular interest to science. They normally consist of one or more lenses or mirrors that force light beams to change course, forming whatever image of the world our needs dictate. Telescopes draw our gaze to the furthest reaches of space, far from the sights that dominate our eyes on a daily basis. Microscopes, on the other hand, turn our sight inwards, to the minute aspects of an inner space. So both instruments opened the gates to inaccessible worlds, making an indispensable contribution to the progress of science.

Though the telescope appears to have been invented in Holland, Galileo was the first person to turn his simple refractor to the skies. "But without paying attention to its [the telescope's] use for terrestrial objects, I betook myself to observations of the heavenly bodies (Fig. 6.1). First of all, I viewed the moon as near as if it were scarcely two radii of the earth distant. After the moon, I frequently observed other heavenly bodies, both fixed stars and planets…"[10] In its basic form, the refracting telescope consists of two lenses. The large, objective lens gathers and bends light rays from faint objects and focuses them at the end of a suitable tube; the image formed is magnified by the ocular lens or eyepiece, bringing an entire, far-off world close up.

Fig. 6.1 Galileo displaying his telescope to Doge Leonardo Donato in 1609, as imagined c. 1900 by Henri-Julien Detouche (Wikimedia commons/PD)

A second type of telescope is the reflector, in which the incoming beam is concentrated by a large concave mirror, and then bounced off smaller mirrors before reaching the eye. Though Newton was the first person to use such a telescope, modern-day reflectors are enormous. For instance, the LBT (Large Binocular Telescope) constructed in Arizona consists of six large mirrors, each of which is almost two metres across. Regardless of type, the size of a telescope is of crucial importance, since it determines the light rays that can be gathered. But because mirrors are easier and cheaper to make than lenses, most large aperture telescopes are reflectors. Use of telescopes that allow for combinations of lenses and mirrors is also increasing.

Nowadays, at every point on the planet, on plains and mountains, in countries rich or poor, alone or in combinations, watchful eyes scan the boundless world of the stars and planets. Indeed, the telescopes or observatories dotted over the expanse of our precious planet are not restricted to tracking visible light. Also important is the role of radio telescopes first developed in around 1960, which investigate the entire electromagnetic spectrum beyond visible light. That is how a new field, that of radio astronomy, has substantially enriched our knowledge of the universe and frequently offered scientists unanticipated discoveries. One important moment, for instance, was the pinpointing of the first neutron star or pulsar; to date, over three thousand of these strong radio sources have been recorded by radio telescopes.

Of the many such telescopes and observatories scattered around our planet, two are worth special mention. The first is the Arecibo Observatory in Puerto Rico, with a reflector diameter of around 305 m. It was an ingenious construction that dominated astronomical research for many decades, especially in the search for extraterrestrial life. But Arecibo's brilliant career was destined to come to a dramatic end. In December 2020, local hurricanes led to successive cable failures that brought its instrument platform crashing down from 150 m, causing irreparable damage to the enormous reflector. Arecibo had lost the world's top spot as the largest reflector a few years earlier, since the Chinese FAST radio telescope, which began operation in 2016, has a 500-m reflector (hence the acronym Five-hundred-metre Aperture Spherical Radio Telescope). Yet despite its impressive capabilities, FAST has so far been limited to discovering new pulsars, as well as detecting a signal that was believed to be of potential extraterrestrial origin. The enthusiasm that followed soon died down, however, as the signal proved to be due to terrestrial interference.

One impressive development over recent decades has involved large telescopes placed in orbit above the Earth's atmosphere, so as to avoid its side-effects. These devices mainly track infrared and ultraviolet forms of radiation such as X-rays, all of which are sisters of visible light and carry important information from the unseen life of stars and galaxies. In 1990 an ambitious space telescope named for the leading astronomer Edwin Hubble was placed in orbit 610 km above the Earth (Fig. 6.2). It has a mirror aperture of close on 2.5 m, and is also equipped with delicate scientific instruments. The mirror on the Hubble initially had serious design faults, and the human mission to correct them at that altitude stands as a feat of technical skill and courage. Ever since then the space telescope has fully vindicated scientists' expectations and justified its cost. It has sent incredible photographs from the vastness of space to Earth, detected heavenly bodies as far as billions of light years away, and made a decisive contribution to the cosmological revolution in our times.

It is worth noting that, contrary to popular belief, modern astronomers rarely watch the sky through their telescopes. Images of stars and galaxies used

Fig. 6.2 The Hubble Space Telescope (HST) during the Space Telescope Servicing 61 flight (1993). The new solar arrays are seen here from the aft flight deck, backlit against the black background of space (NSSDCA/Copyright free)

to be recorded on photographic plates, which have given way to a kind of digital camera capable of detecting even very faint stars. Data are stored and analysed with the aid of computers, which can also produce optical images of astral bodies on their monitors.

Turning now to the other end of the universe, the microcosm was explored thanks to the existence of the microscope. It too was invented in Holland, and is attributed to spectacle maker Zacharias Janssen. As early as 1665, leading English physicist Robert Hooke published a major treatise entitled *Micrographia*, describing and depicting flies and fleas, as well as details of plants. Hooke used compound two and three lens microscopes, illuminating his specimens with oil lamps.

In modern microscopes, the use of achromatic lenses and developments in lighting techniques have led to impressive results. Objects can be magnified up to 2,500 times, while microscope resolution enables characteristic details to emerge. Research was greatly boosted by the electron microscope, which uses an electron beam in place of light rays. The beam is focused with the aid of electrical or magnetic fields, forming the magnified image of the object on a fluorescent screen. Like visible light, electrons exhibit a dual nature as particles and waves, but are of much shorter wavelength. Electron microscopes are thus indispensable when it comes to depicting microscopic objects, magnifying up to a million times.

That being said, the development of quantum physics and technological progress have led microscope technique to even more incredible achievements. The scanning tunnelling microscope uses a metal probe with an extremely fine tip. A beam of electrons tunnelling between the probe and surface of the sample is even capable of depicting the images of atoms or molecules on a computer screen. A similar technique is also used in the most advanced kind of modern optical microscope, which has "photon" scanning and offers ten times the resolution of conventional models. In the latter case it is photons rather than electrons that run through the probe, and the microscopes are suitable for observing biological entities such as viruses and microbes.

Over recent decades the control and use of light has progressed by leaps and bounds. Indeed, some scientists are bold enough to predict that a whole new branch of applications known as photonics will oust electronics from the forefront of our technological era.

6.5   Light in Creative Captivity

The use of telescopes and microscopes is mainly restricted to science. Yet the opportunities offered by capturing light opened up impressive new horizons in entertainment as well as in art. The camera is a characteristic example. Its roots lie in the *camera obscura* described in detail by Leonardo da Vinci, which projected inverted images of the outside world via a small opening. Portable cameras obscura were also built later on, with all the basic mechanisms in a modern camera, e.g. lenses and a focusing function, but could not record the images formed. That major step was made in the nineteenth century, when Frenchman Louis Daguerre and Englishman Henry Fox Talbot pioneered photographic technique. All the same, the world's first photograph was apparently taken by a French doctor named Nicéphore Niépce, who focused a nature scene on tin leaf coated with light-sensitive bitumen.

For many years, modern photographic technique relied on the fact that certain compounds of silver used to coat film are sensitive to light. The visible image captured by a camera was recorded due to changes in chemical structure. As Roland Barthes comments: "It is often said that it was the painters who invented Photography (by bequeathing it their framing, the Albertian perspective, and the optic of the *camera obscura*). I say: no, it was the chemists. For the *noeme* "That-has-been" was possible only on the day when a scientific circumstance (the discovery that silver halogens were sensitive to light) made it possible to recover and print directly the luminous rays emitted by a variously lighted object."[11]

In the pre-digital era, the next stage was to develop the film, so-called "fixing" to make the image permanent, and printing. Pre-digital cameras themselves consist of a system of lenses which focuses any light beams reflecting from the object onto the film. They also have a diaphragm to control the amount of light and usually also the depth of field, while a shutter controls exposure time.

Nowadays, our familiarity with cameras means this bare outline description will suffice. We know that all sorts of paraphernalia are associated with using them: lighting accessories for night photography, wide angle and telephoto lenses, light meters and filters. In the early twentieth century, colour photography opened up new horizons in the compelling depiction of objects and moments in our world.

With the continuous improvements to equipment and methods brought about by the march of technology, the fact is that photography has entered every sphere of activity in modern culture. It has become a substantial player in everyday life, transmitting information with clarity and authenticity, and expanding knowledge in areas inaccessible to the eye. The claim that photography is perhaps the most important invention since printing is no exaggeration. In its great moments, with the artist's inspiration always uppermost, photography has been transformed into an indisputable art form.

Digital photography stands as one radical development that has overturned the foundations of photographic technique so far, creating incredible possibilities for a future which is already here. It is something of a paradox that the invention of conventional film marked a turning point in the history of photography, the significance of which is now apparently to be rivalled by its disappearance. Instead of film, the heart of a digital camera is a light-sensitive integrated circuit, the charge-coupled device or CCD for short. Once light has entered the lens and reached the CCD, the camera's image processor takes over. Information from the incoming light is recorded in millions of pixels, which form the basis for composing the end image. One major advantage of digital cameras is that users can see how their pictures turn out instantly, on a colour screen, and so can delete unsatisfactory photos. Image processing can also be done by PC, while storing photos is easier and more convenient.

The advantages of digital photography have thus almost entirely displaced conventional cameras even in the domain of art. In fact, since modern mobile phones are equipped with digital cameras, the worst has pretty much come to the worst. Everyone in the country can claim to be a photographer, and not just social media but also television and the newspapers are inundated with photographs of varying quality and content. The obvious victim is usually the private life of ordinary citizens, which is seen as vital to protect. Even so, the threats against it are proliferating as technology progresses.

Technical developments aside, capturing light will always remain the basic principle of photography. The limitless forces and capabilities arising from that capture make light resemble a fairytale genie. Imprisoned in a bottle, it too promises that if set free it will place all its magical powers in its master's service.

One natural, admirable outgrowth of photography was the cinema. Its invention is credited to the Lumière brothers, who held the first public screening in 1895, charging admission for ten one-minute films. In the cinema, imprisoned light is used to record and then render real movement, so by necessity recording machines (movie cameras) rapidly diverged from projectors. The famed celluloid that is exposed to light is both durable and elastic, making it easy to wind onto projection reels. What makes it possible to show events in constant motion is the inertia of the human eye: isolated pictures shown in quick succession create the illusion of continuous flow.

Initially an intriguing means of entertainment, cinema rapidly revealed its enormous commercial and artistic potential. Immediately after World War I, Charlie Chaplin, Buster Keaton, Fritz Lang and Sergei Eisenstein—to mention but a few of the best-known names—were to set their seal on the silent cinema with their masterpieces. But full colour talkies were not long in coming. Constant technical improvements, special effects, picture enlargement and widening methods rendered the illusion of reality ever better. This led the cinema to an unprecedented boom after World War II, making it a popular medium not just for entertainment, but also for social criticism. "Cinema," notes Luis Buñuel, "acts directly upon the viewer in presenting concrete beings and things, isolating him in silence and darkness from what we might call his normal 'psychic habitat'. For that reason film can captivate him like no other form of human expression. But it can also dull him like no other."[12]

It is true that as an art form, the cinema has one peculiarity to it. A host of factors such as the acting, the script, the photography and the music have to come together for the final aesthetic result. The films of Ingmar Bergman, Carl Dreyer, Federico Fellini, Luis Buñuel, Andrei Tarkovsky, Luchino Visconti, Akira Kurosawa and so many others opened up new horizons in mankind's relationship with the world and the courses charted by the psyche, highlighting moments in history and life in an unparalleled manner. As Andrei Tarkovsky stresses: "The director's task is to recreate life: its movement, its contradictions, its dynamic and conflicts. It is his duty to reveal every iota of the truth he has seen – even if not everyone finds that truth acceptable."[13]

It goes without saying that one of the most important factors accentuating a good cinematic image is light, and the inspired use of its pathways on faces and landscapes. As Fellini characteristically notes, "The heart of every object, both in cinema and in painting, is light. In cinema the light comes before the subject, the plot, the characters: It is light that expresses what the director means."[14] This quest calls for expert knowledge of both the techniques and the aesthetics of light. Right from the shooting stage, artificial and natural light, filters and camera angles are used to highlight the actors and the details of space or, if required, to keep them in the warmth of half-light, in tough coexistence with the shadows.

Yet while the object of photography is an authentic hint at what is visible and invisible, the cinema has the added difficulty of highlighting motion, the collusion of light with the ever moving or the stationary. The magic of the cinema lies in the fact that once in the dark auditorium, viewers are called upon to converse with the actors on the screen and with themselves.

Nowadays, advances in computers and digital techniques often edge the cinema into the realm of simulation and virtual reality. And though the results are often admirable, they can hardly approach the special magic and quality characterising the finest moments in cinematographic art.

It is nonetheless a fact that the gadgets and human inventions that tame or imprison light, like the cinema, are seemingly without end. Cameras or telescopes, glasses to improve sight, projectors and television cameras cater for our everyday needs, sometimes serving special purposes. But what needs to be stressed, and emphatically so, is that over the long course in which man has tamed light ever more, we have never seen light itself; its presence always remains invisible. Yet the magical and invisible entity we call light has proved capable of changing human life and its prospects, of recording history, and of leading to new forms of self-knowledge and culture.

6.6   Light Collaborates in the Microcosm

Recent decades have seen a new light source make its timid appearance, before rapidly gaining an important place in science and our life: laser beams. The term laser is an acronym of the words Light Amplification by Stimulated Emission of Radiation. Hidden behind the obscure meaning of that phrase is a simple yet revolutionary principle of physics, linked to the so-called coherence of light energy. The same principle governs the use of semiconductors and fibre optics; in other words, it is a substantial factor in today's technological revolution.

Impressively enough, the idea of stimulated emission which led to the laser and its applications can be traced back to a 1917 publication by Albert Einstein. The ingenious physicist had worked intensively on the stimulation and emission of radiation from atoms. "A splendid light has dawned on me," he enthused. So the man who hallmarked our era with the breadth and depth of his theoretical thinking paved the way for two revolutionary applications in physics: nuclear energy and laser beams.

What, then, was the revolutionary idea first conceived by Einstein? Let's take a slightly closer look at how ordinary light is produced in a fluorescent lamp. The atoms in the tube—in a gaseous state—are initially at rest. In the absence of any external stimulus, the electrons going round the nuclei are at the lowest energy level. But if the light is supplied with electricity, some of the electrons absorb energy, become excited and move at greater distances from the nucleus. This excited state does not last long: in less than a billionth of a second the electrons revert to the ground state, emitting the energy they have gained in the form of light. Emission is spontaneous, the photons move in random directions and the radiation produced is relatively weak.

If, however, a photon of the right energy level finds itself near the atom while the electron is excited, then things take a different course. Due to the presence of the photon, there is a very good chance the electron will be induced to drop down to a lower level. Only this time it will emit a photon that is an identical copy of the one that caused it to drop down! The two photons—the original one and that produced by the stimulated transition—have the same directionality, energy and phase. It is not possible to establish which of them caused the emission and which was generated during the transition.

Indeed, the effect may not end there. Thus the twin photons may bring about stimulated emission in other atoms, resulting in the production of new photons, always at the same frequency and in the same direction. It is obvious that if we can harness this process, which snowballs to constantly produce identical photons, it will end in a light beam with highly desirable characteristics: the beam is monochromatic and highly directional. Indeed, in contrast to the photons produced by excited atoms in spontaneous transition, the photons in stimulated emission have a fixed phase relationship, thus emerging from the emission as "coherent" radiation. This is a key concept, for lasers differ from normal light as much as music does from noise. Ordinary light is emitted from its source—be it the sun, an electric light bulb or fire—in a range of wavelengths heading in all directions. On the other hand, laser light consists almost exclusively of one wavelength. Its waves move in the same direction, amplifying each other like the voices in a well-trained choir.

There is, however, one vital precondition to creating this coherent photon beam. There obviously has to be a large number of atoms in an excited state for the stimulated emission to be effective. And this is where the main difficulty lies. Although stimulated emission is a natural process in atomic systems, it is not in the least normal for a lot of atoms to be in a high energy state. The laws of quantum mechanics demand otherwise. Most atoms are in the ground state, a few are on the first level, and fewer still on higher energy levels.

So for stimulated emission to be of some use, the atoms have to undergo so-called "population inversion". In other words, a state outside their normal equilibrium, where most of the electrons will be excited. Using technology developed for the purpose, this population inversion can be achieved via "pumping" in one of several ways: using special valves, by electrical discharges or by chemical reactions. So depending of the type of pumping or how they operate, as well as on the primary material used to generate the beam, there are many kinds of laser, and new ones are constantly being developed.

These types use a range of materials, including solids and gases. Consequently, laser beams exhibit different characteristic properties in each case. As their main property, wavelength varies from the ultraviolet through visible light to the infrared. Even "blue" lasers, which are extremely difficult to build but have distinct advantages, are already being used in the audiovisual industry and in computers.

Synthetic ruby—a transparent crystal of aluminium oxide—was the material used by the American physicist T. Maiman to build the first laser. Despite being low-strength, helium–neon lasers are now very widespread. High voltage is applied to the ends of a tube containing a mixture of the two gases. As if there were some love chemistry between them, the electrical discharge then occurring in the helium puts the neon atoms into prolonged stimulation. The radiation emitted, corresponding to the red or infrared area of the spectrum, is thus due to the neon.

As new applications are constantly being found for lasers, the types they come in have also increased. In addition to the crystal and gas versions already mentioned, new types have been added: chemical lasers, glass lasers and semiconductor lasers. The last of these harness the population inversion created between the conduction band and the valence band in a semiconductor. A typical example of this category is the gallium arsenide laser, which was also the first semiconductor version developed.

Regardless of their type, an enormous number of laser devices now serve mankind. Some are complex affairs as large as a football pitch, whereas others—such as those that read CDs—are even smaller than a grain of sand. Some laser devices emit pulses lasting billionths of a second, while others can emit constantly for decades. And as for their strength, some lasers can focus light so bright that it can vaporise any material on earth.

Since there is some confusion on the matter (only added to by science fiction films), it is worth stressing that lasers are not energy sources, nor can they give more energy than they take. Instead, they are minimally efficient energy converters, which use the stimulated emission process to generate high quality, versatile energy. Like a bow, which stores up energy to fire an arrow when released, lasers concentrate energy and release it as radiation of one wavelength alone, which propagates in one direction alone. By collaborating in the microcosm, light has once again revealed both its fascinating nature and its limitless capabilities.

6.7   Light That Burns and Light That Creates

Humans have often stumbled across some major secret of nature while doing science, but the road is rugged and demanding before the first useful applications of it appear. It took several decades and the contribution of a delicate technology for the collaboration between light and the microcosm to blossom into so many applications. Today, uses for lasers really do appear endless; and nor is it possible to predict where they may yet be used in the future.

The large-scale development of lasers is due to their characteristic features: a coherent, monochromatic beam, high directionality and focusability, and high instantaneous power in certain types. The intensity of a laser beam is of particular interest in industrial applications, and can often be incredibly powerful. On a hot day, the energy from the sun that reaches your finger is no more than a few dozen watts, but on a piece of steel the same area, a powerful industrial laser can easily concentrate ten billion watts!

In the simplest applications, a laser beam is used as an unbending straight line, offering precision and speed in mechanical tasks, construction or surveying. Lasers can also be used to ensure that underground or undersea tunnels are perfectly aligned when being dug. Beam directionality has even enabled us to measure the distance from the Earth to the Moon to within a few centimetres. To that end, astronauts on the Apollo 11 mission placed an array of reflectors on the moon, to bounce back a laser beam sent from Earth. Measuring the distance precisely has yielded useful information on our planet's geological history. Lasers also assist in monitoring continental drift, which is of major significance in understanding and possibly even predicting earthquakes in the future.

The development of lasers has benefited many branches of science and created entirely new ones. For instance, spectroscopic measurements of value across a broad range of scientific activities have become exceptionally accurate. Spectacular strides have also been made in holography, enabling the reproduction of 3D images. Holograms have all sorts of uses, from artwork imaging and teaching to fraud prevention on credit cards.

One further high-tech application for lasers is in telecommunications. Unlike radio waves, with wavelengths extending over hundreds of metres, or television waves, which have wavelengths of several centimetres, laser beams are ultrashort waves. As a result, they are capable of carrying vast numbers of messages concentrated in a narrow band of frequencies. New horizons have thus been opened up in telecommunications by the beams, which either pass through the atmosphere or are directed along optical fibres. The fibres are thousands of glass strands as fine as a human hair, bundled together to form a kind of light tube as flexible as a cable. An internal coating reflects the light, enabling its energy to be transmitted over long distances at minimal loss.

As was only to be expected, lasers were not long in spreading to more humdrum everyday things. Nearly everything on sale nowadays has a barcode on it. Laser light scanners can read the code's bars and spaces, and convert them into an electrical signal, transferring information on a product's ID to a computer database that keeps details of its price and availability. A similar method is used to monitor pollution, the true bane of modern megacities. As they absorb light at particular wavelengths, atmospheric pollutants leave a kind of fingerprint on a reflected laser beam. Data from this can then be instantaneously analysed by computer to generate a table of pollutant values.

The revolution that the Compact Disc or CD brought to how we enjoy music was also largely due to lasers. An indestructible optical needle in the form of a narrow laser beam reads the musical information stored in digital form on the discs. However much vinyl fans may (quite rightly) complain that something of the mellowness in the music is lost, the convenience and durability of CDs led to them spreading so rapidly, before they in turn were replaced by digital audio files stored directly on devices. Such is the march of technology.

Like a magic wand, the beneficial powers of lasers have also touched human pain. There are already numerous uses for laser beams both in surgery and for various other therapeutic purposes. In eye surgery, for instance, high intensity beams are used to weld the retina into position without an incision being made. Even advanced myopia can be overcome without lenses, simply by using the high-precision intangible scalpel provided by a narrow laser beam.

Art also numbers among those human activities where lasers have opened up new horizons or exerted an influence. Of course, many of the shows that use holograms or intense laser beams can hardly be described as artistic creations in their own right. As is almost always the case with technology, the use of lasers in art can only be thought of as supplementary; inspiration and a sense of culture remain the artist's primary weapons, or at least they should be. On the other hand, much is already being gained from lasers in so-called cultural heritage, which usually suffers at the hands of its beneficiaries. Thus beams can be used to clean, scan or reproduce artworks from bygone eras. Laser specialist John Asmus summed things up well when he said: "People think of the laser as a mere device, but that's a narrow perspective. We're learning the potential of light from a profoundly powerful technology."[15]

It is obvious that in addition to their beneficial side, laser beams could hardly have avoided being pressed into the service of evil. "At many things—wonders, terrors—we feel awe, but at nothing more than at man," as Sophocles said so long ago in his *Antigone,* before going on to stress that despite displaying wisdom and intelligence in the arts, humans were led "both to evil and to good."[16] Though limited until recently, military applications for lasers (in the service of major world powers, naturally) appear to have strong prospects. Early on in Vietnam, as well as in the Iraq War, the remote control bombs and missiles fired by the Americans used laser beams to pick out their target and navigate accordingly.

The constant expansion in uses for the laser—the light that burns, communicates, reads or creates—justifies the gift Prometheus gave us, while also broadening the ancient myth's symbolism. Prometheus not only stole fire for the sake of mankind, he also taught us how to use light. It was the great gift he gave our life and civilisation.

6.8   Light's Wanderings in Art

Any book purporting to have the investigation of light as its basic aim is duty bound to include some reference to art. There can be no doubt that both in the act of creation itself and in showcasing art, light is a fundamental, existential constituent.

The relationship between art and light stands in itself as an entire, inexhaustible subject. In the final analysis, art is a magical encounter between light and the artist's inspiration.

In art, then, light's collusion with figures and space is glorified; art preserves the memories of light and its acrobatics; art forces light into constant explosions and doublings back. In the various dimensions of art, an invisible light reveals its invisible charm, as well as the negation of that charm.

Of course, we have no need of witnesses. Let's simply note that photography and the cinema, two modern expressions of art, rely on light to record or narrate their stories. Sculpture and architecture on the other hand, which have had unique masterpieces to boast of since ancient times, cannot be conceived of without light's presence and caresses. One only has to observe the way Cycladic figurines breathe in light, or the infinite playing of light on the columns of the Parthenon. Besides, the materials used by contemporary sculpture, whether marble, metal or wood, reveal their own identity in light. The works of Henry Moore or Brancusi, Rodin and Giacometti, Arp and Calder draw strength from light so as to exist (Fig. 6.3), rendering that strength many times over.

Greek light, "angelic and black" as the poet George Seferis put it, lends a special dimension to Modern Greek sculpture. Yannoulis Chalepas, Manolis Tzombanakis and so many others have had to struggle with a pellucid, absolute light that brings things to the fore, but is also unforgiving.

In a book on this subject, Gerasimos Sklavos deserves pride of place. A tragic figure in Greek sculpture, he made a universal faith of his struggle with light. "Light," he notes, "is the beginning of the universe, the beginning of creation. What is the Soul? It is light. Light-Motion-Life, Soul-Light-Creation."[17] Having studied at the Athens School of Fine Arts, in 1957 Sklavos settled in Paris, where he was to combine traditional values in sculpture with new modes of expression. Subversive and original, he believed that sculpture should collaborate with science; and he appears to have had a very firm grasp of the laws of physics. He had even patented an invention to do with processing a certain very hard rock by fire. As he himself once wrote, "The laws governing the genesis of rocks and the genesis of the universe have often concerned me; I figured that both occurred under the influence of

Fig. 6.3 Rodin's *Beside the Sea* (c. 1907), Metropolitan Museum of Art (CC0)

enormous heat."[18] At the time, the idea that the universe had a superheated beginning was virtually unheard of!

Sklavos' experiential, charismatic relationship with light extends over a wide-ranging oeuvre not excluding paintings. A variety of materials are used in his sculptures: plaster, for instance, in *Icarus*; Egyptian pink porphyry in *The Escape of Matter—Interplanetary Light*; welded iron in *Ray*; Pentelic marble in *The Final Insight* and *The Beginning of the World*. Yet regardless of each sculpture's material and form, what characterises Sklavos is an obsession with the metaphysics of light. "Light takes the lead," he proclaims.

This peculiar relationship with light reaches its apotheosis in Sklavos' magnificent work *Delphic Light* (Fig. 6.4). He seems to have been fatefully attracted to the sacred landscape at Delphi, which exudes an all-powerful sense of intellectual light. "At Delphi my soul ignited," he writes in surviving notes. That "ignition" was channelled into the sculpture he created, working the marble as the wind would have shaped it down the course of millennia. The piece appears suspended in light and space. All sides are prominent, while its form brings to mind a multidimensional structure floating in space, only momentarily touching the ground.

Gerasimos Sklavos, the "sculptor of light", was to meet a tragic end. In January 1967, his studio had been plunged into darkness by a short-circuit when a large granite work characteristically entitled *The Girlfriend Who Wouldn't Stay* fell on the artist, costing him his life. At the time he was already

Fig. 6.4 *Delphic Light* (1966) by Gerasimos Sklavos, Delphi, Greece (Courtesy of Dimitrios Grekas)

tasting international acclaim. It was as if his end—an end in light—wanted to showcase his fateful love affair with his materials and light.

In painting too, artists are called upon to carry out their intentions by means of light. Of course, the role of colours is paramount here. Contrary to a popular fallacy, possibly due to the large number of excellent sculptures that have survived, painting did not lag behind in ancient times. As the celebrated Greek painter Yannis Tsarouchis underlines, "The little knowledge we do have of ancient Greek painting not only enables us to see that painting did exist in Ancient Greece, but that it was the greatest, most unique, perhaps the only painting."[19] Luckily enough, a whole archive of information on Ancient Greek art was preserved by Pliny the Elder in his monumental *Natural History*. Book XXXV is on Ancient Greek art, where, in addition to an infinite wealth of colours, one encounters incredible technical dexterity in preparing them. "Eventually," Pliny says, "art differentiated itself, and discovered light and shade, contrast of colours heightening their effect reciprocally. […] Some colours are sombre and some brilliant, the difference being due to the nature of the substances or to their mixture."[20] As we go on to learn of woad and ochre, Tyrian purple and indigo, crimson and cinnabar, the magic of colours leaps out from the pages of Pliny to reach the present day.

It is a fact, then, that from cave art to the amazing wall paintings of Thera and Crete, and from the otherworldly figures of El Greco (Fig. 6.5) to the unrivalled sensation created by the paintings of Magritte, Hieronymus Bosch or Picasso, colours suggest, synthesize and surprise. In Rembrandt's art, the handling of colour and light are just as important as the subject.

We have no need of further witnesses: one characteristic fact is that entire periods of painting have had the world of colour as their reference point. One revolutionary movement that was to appear in late nineteenth century France, and was at first mockingly called 'impressionism', proclaimed the need for direct contact with nature, and the liberation of painting from traditional studio conventions. Leading artists of the time such as Manet, Renoir, Monet and Degas formed part of the movement, which placed emphasis on the changes in light, as well as on contrasts or the intensity of colours.

Furthermore, the early twentieth century saw the growth of Expressionism as another powerful artistic movement, mainly in Central and Northern Europe. Here the characteristic feature was the distortion of reality so as to express the artist's internal perspective. Expressionism proclaimed the end of everything old and the beginning of a new era, and its influence was to spread to literature, music and even the theatre. Van Gogh is regarded as the forerunner of Expressionism in painting; its exponents include Cézanne, Kandinsky, Kokoschka and Chagall. It should of course be stressed that

Fig. 6.5 An otherworldly landscape and figures in El Greco's *View of Mt. Sinai and the Monastery of St. Catherine* (1570) (Historical Museum of Crete, courtesy of the A. & M. Kalokairinos Foundation)

whether in art or politics, a "movement" cannot always be governed by common rules, nor is it homogeneous. With that necessary footnote in mind, in Expressionist works the need for expression is served by unusual and at times unreal colours, while inner passion and emotion predominate.

Lastly, one other twentieth century artistic movement whose representatives were characteristically labelled *fauves*—wild beasts—regarded colour as a painting's main feature, and treated it with true adoration. With Matisse as one of their brilliant exponents, artists here attempted to express their feelings through an explosion of pure colours, arranged independently in each painting. Of course, at the same time Cubism was overturning perception in visual art and introducing new ways of structuring aesthetic objects. As Nobel Prize-winning Greek poet Odysseas Elytis pithily noted, "You can never depict the sun, you can only reintegrate it into the nature of things. The Impressionists caught a moment in the air, but Cézanne's investigations soon came to displace and ultimately overturn the significance of that feat. With

Cubism, light turned from a fleeting impression into a structural feature. And the Fauvists turned it from a cause into a result."[21]

There is of course no need to invoke the history of art for the role of light and colours in painting to clearly emerge. That role is both self-evident and dominant, manifesting itself right next to us, in exhibitions and galleries, art museums and—for those lucky enough—on the walls at home. Gently or powerfully, with tones sometimes subdued and sometimes brilliant, familiar and sometimes unrealistic, colours may not determine an artist's inspiration, but they daringly and persuasively serve it. At the same time, however, it is as if they hint at their common origin from a select totality: the single, indivisible nature of light. That fiery chariot, which constantly returns and is analysed into colours, touches human beings and creates the uniqueness of their art.

That being said, it is obvious that any attempt to make brief reference to the role or function of light in art is doomed to be ineffective. At the very least it would call for yet another "autobiography of light", but one with a different style and contents. So the preceding pages can only be seen as pinpricks of light, or tiny hints. Light's presence in art appears to know to bounds. Any attempts to define it recalled the author's attempt to apprehend light in her face, in the movements of her hands or in her exquisite eyes. But for his part, the author was not slow to realise how futile the exercise was, and to further reject any interpretation or description. So, all that remained was to admire the chiaroscuro and the outburst of colours that surrounded her wanderings.

7

The Universe: The Empire of Light

7.1 Light Betrays the Secret of the Galaxies

So far, we have seen light's commanding presence in scientific developments and the creation of life, its acrobatics in the quantum microcosm and the thousand faces of its everyday manifestations. Light appears dominant everywhere, its domain extending far beyond the Earth. The universe itself is the empire of light.

A century ago, the general conviction was that the universe began and ended in our galaxy. The stars even appeared to be concentrated in a hazy glowing band, which folk beliefs associated with the River Jordan or scattered straw. Italo Calvino's Mr. Palomar—a clear reference to the Caltech observatory—puts it eloquently when he says: "To a large extent the sky is streaked with light stripes and patches; in August the Milky Way assumes a dense consistency and you would say it is overflowing its bed; the dark and the light are so mixed that they prevent the effect of perspective of a black abyss against whose empty remoteness the stars stand out, in relief; everything remains on the same plane: glittery and silvery cloud and shadows."[1]

Belief in the uniqueness of our galaxy was overturned in 1924, when the leading astrophysicist Edwin Hubble proved that the distant points of light picked up by telescopes were in fact galaxies, too. Immanuel Kant had suspected the truth of this as early as the eighteenth century, and had even called the distant nebulae "island universes", separated from us by vast empty space. We now know that far from being unique, our galaxy is just one ordinary instance among billions of others.

© The Author(s), under exclusive license to Springer Nature
Switzerland AG 2024
G. Grammatikakis, *The Autobiography of Light*,
https://doi.org/10.1007/978-3-031-56917-3_7

Galaxies are the building blocks of the universe, and are in essence enormous systems of stars. Tremendous distances separate the stars or "cells" in each galaxy. Their number usually runs into the hundreds of billions, while interstellar space is filled with gases and dust.

There are galaxies similar to our own and other spiral-shaped ones; elliptical galaxies and exceptionally bright ones—the mysterious quasars— traceable in the furthest reaches of the universe. Our nearest galaxy is the Large Magellanic Cloud, about 160,000 light years away. Our neighbourhood also includes the Andromeda Galaxy (Fig. 7.1), faintly visible even to the naked eye, which is two million light years away.

As is true of each individual star, analysing the light that a telescope receives from a distant galaxy yields valuable information on its movements and chemical composition. It also led to an important, unanticipated discovery made yet again by Hubble, when he was closely studying the spectra from galaxies.

One phenomenon familiar to our experience, relating to sound, can serve as a first step in understanding this discovery. We have all noticed how the

Fig. 7.1 The Andromeda Galaxy (Makis Palaiologou, courtesy of FORTH Institute of Astrophysics, Greece)

frequency of a sound changes when the source producing it moves. An ambulance siren, for instance, sounds higher and higher pitched as it approaches it us, but then deepens (i.e. its frequency decreases) as soon as the ambulance heads past us. The same effect is noticeable with train whistles. It was often recorded in old westerns, since trains and train chases by outlaws were often the centrepiece of the action. As it approached the heroes or the station, the train whistle shrieked higher and tension mounted; but as it drew away, the whistle faded and its frequency decreased. In other words, it sounded deeper.

This change in sound wave frequency depending on how a source moves has long been known to physics. It is called the Doppler effect, after the person who discovered it. If the change in frequency is measured, a simple mathematical relationship can then be used to calculate the speed at which the source is moving. The interesting thing is that the same effect is also seen with electromagnetic waves. When applied to radar waves it is used to clock careless drivers. Since the waves bounce off cars, the change in their frequency gives their speed, and whether or not they are over the limit.

Things don't essentially change when it comes to visible light, which is also an electromagnetic wave; but they do have more spectacular consequences. As a light source recedes, the frequency of the light reaching our eyes decreases. In other words, its wavelength increases. A typical example involves a stationary source emitting violet light. Its frequency will be shifted to a longer wavelength—to cyan, let's say—the moment the source starts to move away. Naturally enough, the opposite will be seen if the light source is approaching at high speed, in which case violet may become ultraviolet or even shorter wavelength radiation. How much the frequency shifts depends in general terms on the speed of the light source.

With that necessary introduction over, let's get back to starlight and the galaxies. Having confirmed the existence of other galaxies, Hubble set about classifying them in terms of distance and studying their light spectra. As was to be expected, when analysed using special equipment they resembled the spectra of stars in our own galaxy, in each instance also indicating the star's chemical composition. But there was one substantial difference: the spectra derived from other galaxies were, with just a few exceptions, shifted towards red.

The conclusion to emerge from these persistent observations was sensational: the galaxies were receding. What's more, they were doing so at high speed. That was why they showed a corresponding shift to longer wavelengths. Their shift magnitude was in fact directly proportional to the distance from the galaxy. In short, the further away a galaxy was, the greater its recession speed. This was expressed in a simple law formulated by Hubble, which has quite rightly borne his name ever since. The actual recession speed of galaxies itself can reach incredible levels. Our near neighbour, the Virgo Cluster, is receding at 11,200 km per second, but speeds ten times as fast have also been recorded. The growth of radio astronomy, which intercepts radio waves from stars, has revealed that the speeds of distant quasars may even approach that of light!

This is where one initially naive-sounding question arises: what are the galaxies receding from? The human-centred view of the cosmos that was dominant for so many centuries gives the obvious answer that they are receding from us, from our own galaxy. But that runs counter to a basic principle of cosmology, which states that the universe is homogeneous, and consequently attaches no particular importance to our galaxy. In reality, neither is our galaxy at the centre of the universe nor indeed does any such centre exist. In earlier times Epimenides of Knossos had ruled out the existence of any "navel of the Cosmos", neatly arguing that if there was one, then it would be visible to gods but not to mortals.

So the galaxies are not just receding from our own Milky Way, but each one is moving away from all of the others. If they exist, the inhabitants of a distant galaxy will be awestruck to see the others receding, in accordance with the same simple law formulated on insignificant planet Earth by Edwin Hubble. As a matter of fact, it's worth noting that Hubble had studied law, and even practised it for a time. His father was in the same profession, and Edwin, born in Missouri in 1889, was the fifth of his seven sons. While continuing to study law at Oxford he excelled as an amateur boxer, and came close to taking up the noble art professionally. Luckily enough, however, from early on he showed interest in the magical world of astronomy. So Hubble gave up law for good in 1914, to be taken on as an assistant at the University of Chicago Yerkes Observatory. There he studied from scratch, and soon earned his doctorate in Astronomy.

As World War I had already broken out, Hubble joined up to serve in France, where he was injured. Fortunately for him and for science, from 1919 until the end of his career—he departed this life in 1953—Hubble was to work in the large, newly-built Mount Wilson Observatory in California. His remarkable contribution to astronomy began with the discovery that the Andromeda Nebula in our celestial neighbourhood was in actual fact a galaxy.

The honour that the scientific community reserved for Hubble was thus entirely justified—as has already been mentioned, his name is proudly borne by the major space telescope launched by NASA and the European Space Agency in 1990, which has been orbiting outside the Earth's atmosphere ever since.

The universe, then, is not static and eternally the same, as was the accepted wisdom before Hubble. Each of the galaxies is receding from all of the others. This fact, which was revealed by their light, leads to an inescapable truth: however unbelievable it may seem, the universe itself is expanding! Its expansion is constantly creating new space around it. Space itself is extending, so the distance between the galaxies is increasing.

The expanding universe is in many ways similar to the surface of an inflating balloon. The galaxies can be represented by coins stuck to the surface. As the balloon inflates, the coins or galaxies draw apart, since the space between them increases. In fact, as each galaxy is surrounded by all the others, it is hardly strange that it sees itself as the centre of the expanding universe. In reality, just like the surface of the balloon, the universe has neither a centre nor outer reaches. Expansion constantly creates new space around it. So questions such as "what is the universe expanding into?" or "What is there beyond the universe?" don't really make much sense.

The perpetual flight of the galaxies as captured in spectral red-shifting was one of the greatest discoveries of the twentieth century. To this day, ordinary people are amazed at the fact that the universe—that inconceivable Everything—is expanding. As Woody Allen's terrified alter-ego Alvy concludes in *Annie Hall*, "The universe is expanding. Well, the universe is everything, and if it's expanding, some day it will break apart and that will be the end of everything."

Even in the world of science, however, the idea of expansion and its consequences were not easily accepted. The historical and epistemological roots of certainty over the static universe ran deep. One characteristic fact is that the expanding universe was a consequence of the general theory of relativity, conceived a decade earlier by the genius of Einstein. Yet he himself refused to accept this fundamental consequence, perhaps because it violated his strong beliefs on the simplicity and aesthetics of the cosmos. Einstein even went as far as to introduce an arbitrary constant into his relativity equations, which, as a helping hand to an uneasy God, would deter the universe from expanding.

Luckily for that same universe, the short-lived yet brilliant Russian mathematician A. Friedmann saw through the mischief while studying the relativity equations in greater depth. Einstein later confessed that it had been the biggest mistake in his life. The expansion of the universe could no longer be doubted.

Was it really a mistake, or perhaps a momentary lapse of confidence in a genius? In recent years, a multitude of astronomical observations using modern means has ascertained that there really is some kind of cosmological constant, and indeed one leading to an unexpected acceleration in the rate at which the universe is expanding!

Yet regardless of such developments, one truth remains undeniable. It was thanks to light that the galaxies revealed the grand secret of their everlasting flight. Just as when, at some time and place beyond recall, the clarity of the morning light betrayed Her fleeting presence, Her enigmatic smile and the everlasting song of Her eyes.

7.2   The Big Bang: A Day Without Yesterday

The expanding universe revealed by the red-shifting of stellar light leads to one inescapable conclusion. The further back we go in time—as in a film running backwards—the smaller the dimensions of the universe are. The galaxies draw ever closer, the stars first touch and then amalgamate into a cosmic sphere, and the temperature constantly rises. So there was a moment in the now distant past that coincided with the beginning of expansion. That point was the birth of the universe itself.

It follows that the universe has not always existed, as was generally believed a few decades ago. It was born at some point in the past, and already has a long prehistory. At the moment of genesis—which was the beginning of time, and the beginning of everything—all of its matter and energy were concentrated in an infinitesimally small cosmic sphere, or primordial cosmic egg. Then, for reasons we do not know and may never learn, an unimaginably large explosion took place. It was the Big Bang, from which today's universe slowly emerged. A comprehensive, detailed theory formulated on that basis has been confirmed by observations, and is now an article of faith for contemporary cosmology. It should be noted that the same conclusion can also be drawn from the equations in the general theory of relativity. They too predict the convergence of all points in space–time in a mathematical anomaly, which is to be identified with Point Zero of the Big Bang.

Obviously enough, the Big Bang or the moment the universe was created defies easy description—not only on the basis of our own experience, but even in scientific terms. Only poetry can occasionally come to our aid, giving its own version of cosmogony through the words of Jorge Luis Borges:

Neither darkness nor chaos. the darkness
requires eyes that see, like sound
and silence require hearing,
and the mirror, the form that populates it.

Neither space nor time. Not even
a divinity that premeditates
the silence before the first
night of time, which will be infinite.
The great river of Heraclitus the Dark
its irrevocable course has not begun,
that from the past flows into the future,
that from oblivion flows into oblivion.
Something that already suffers. Something that implores.
Then world history. Now.[2]

As science would have it, the history of the world began with a Big Bang, a term which surely also covers our ignorance. It certainly wasn't an explosion in space at a particular time, like a shell exploding. The Big Bang itself created time and space, and occurred simultaneously everywhere. Thus in every galaxy that sees the others receding, the impression now gained is that it was at the centre of the explosion. In reality, conceiving of the point where the Bang occurred is not possible, nor is there any centre to the present-day universe.

At any rate, the enormous impetus that the Big Bang lent to primordial matter caused the flight of the galaxies that we see today. The image of the constantly receding galaxies can even lead us back to their initial starting point, allowing us to calculate how much time has elapsed. In other words, we can calculate how old the universe is. On the basis of Hubble's simple law it was initially estimated that the Big Bang occurred 15 million years ago. Constant revisions were made to this estimate over the course of years, whether due to new data or to details in the cosmological dominants that came to dominate from time to time. But in 2021, on the basis of observations made by the Atacama Cosmology Telescope (ACT) in the Chilean desert, plus findings gathered by the European Space Agency's Planck satellite, scientists reached final agreement on the age of the universe. It was fixed at 13.77 billion years, plus or minus 40 million years. Whatever the case may be, it was in its explosive birth—whether by happenstance or necessity—that the history of the universe, its perpetual evolution and its transformations all began.

In its early steps the universe was superheated and unimaginably dense, but over time it thinned out and cooled down. In fact, the stages in its evolution can be reconstructed on the strength of physical theories and observations, and of course given the arrogance of physicists, who have so gone as far as to describe the first billionths of a second after the Big Bang. In general terms, the light nuclei of the elements were formed in the beginning, followed a few hundred thousand years later by the atoms of matter, and the primordial galaxies about a billion years after that. At some point the stars in the galaxies turned on their light by nuclear reactions. And about five billion years ago, one of those stars, the Sun, illuminated a young planet that would much later be called Earth.

So a cataclysmic Big Bang was the beginning of the cosmic drama, which later evolved into galaxies and planets, radiation or energy; and as far as concerns us, into the living world and human beings. At the same time, it was also the creation of the stage—i.e. space and time—on which the drama was to unfold. This fundamental beginning, which can be deduced from general relativity, spares us irritating questions about what happened or existed "beforehand". On the other hand, it does nothing to get in the way of the sarcastic disposition in Italo Calvino's *Cosmicomics:* "Naturally we were all there—*old Qfwfq said*—where else could we have been? Nobody knew then that there could be space. Or time either. [...] Every point of each of us coincided with every point of each of the others in a single point, which is where we all were."[3]

An ironic description of agnostic scientists' feelings is also given by astrophysicist Robert Jastrow: "For the scientist who has lived by his faith in the power of reason, the story ends like a bad dream. He has scaled the mountains of ignorance; he is about to conquer the highest peak; as he pulls himself over the final rock, he is greeted by a band of theologians who have been sitting there for centuries."[4] It's also worth noting that Georges Lemaître, a Belgian astronomer and Catholic priest, had earlier on expressed the view that the universe had its origins in the explosion of a "primordial atom."

All the same, the indisputable fact is that as a branch of science rather than a myth or a religious need, cosmology was actually born of two parallel paths in the twentieth century. Firstly from general relativity, which laid down its theoretical framework, and then from observing the red-shifting of light, which was to reveal that the galaxies were receding. The birth of the universe at some point in the distant yet not infinite past was the inevitable conclusion arising from Hubble's observations.

Over recent decades, space telescopes and radio astronomy have lent new impetus to exploring the universe and its turbulent course. Yet the inflection point had already been reached, once again thanks to the light of the stars. That light revealed that long before the first human heart ever beat, a dazzling cosmic flame had set in motion a complex mechanism, lending pace and existence to the universe. And long before human eyes ever met the light, countless galaxies were already sailing the cosmic ocean and stars were illuminating the cosmic darkness.

Light, then, is not only our daily companion in our search for the beauty in the world and in art. Through its colour transformations on arriving from the stars, it pointed to the universe's big, well-hidden secret: that it had been born at a particular time in the past. In fact, the cataclysmic explosion 13.8 billion years ago that led to its birth left deep tracks in the skies. Those tracks are once again light of a kind—this time a breathtaking, unanticipated fossil of light.

7.3   A Fossil of Light

In our daily life, light is created at every moment and in many different ways. For instance, it sets out from a ceiling lamp and diffuses through the room; or it arrives as solar radiation, giving life and meaning to the world. Light comes from a fire that radiates heat and warms us, and also from the distant stars, having travelled for millions or even billions of years. The light of day or the weak light of night is a divine game, with infinite routes to take; it may set out from the sun or from an artificial source. On its way it may have been reflected from the moon or from the countless particles of dust in the atmosphere.

So it doesn't seem pointless to ask what the most ancient light is, and which radiation can reveal the depths of the world's time and past, if indeed they can be revealed. Besides, we humans like talking about the past. In studying fossils we strive to discover the origins of life and its first forms, and from ancient ruins we draw conclusions about past civilisations. The fact that there is also a kind of fossilized light does sound paradoxical, but it is an undeniable scientific truth.

From every direction in the sky we receive light that has travelled for millennium upon millennium, and century upon century, to reach us as an irrefutable witness to the brilliant, explosive beginning of Everything. That primordial light confirms the hypothesis that the universe has not always existed. It was created at some point in the distant past; and an unimaginably large explosion, the celebrated Big Bang, was its root cause.

Interestingly enough, although the discovery of primordial light was one of the most important moments in the history of science, it occurred more or less by chance. In 1965 Arno Penzias and Robert Wilson, two physicists working for Bell Laboratories in America, were testing out a horn antenna built to track signals from a man-made satellite. This would have been a boring routine task had it not been for an unexpected hitch.

The sensitive antenna seemed to be picking up inexplicable yet regular noise in the microwave range, i.e. somewhere between infrared and radio wave frequencies. Thorough checks ruled out the obvious explanation that the noise was coming from the antenna itself. In fact, some innocent pigeons that had roosted in the antenna fell victim to science when their droppings momentarily came under suspicion as good dielectric insulators! Quite soon, however, it was established that the weak isotropic radiation did not even appear to be coming from any star in our galaxy or any unknown microwave source. It had to be exogalactic in origin, since the antenna picked up the same signal wherever it was turned.

The correct explanation was not long in coming. A few kilometres away, at Princeton University, physicist Robert Dicke and his associates had been bold enough to predict that the superheated state prevailing after the Big Bang must have left behind some radiation much cooler than it originally was, due to the constant expansion of the universe. The Princeton researchers had even designed an antenna to track primordial radiation.

By a cruel whim of fate, it was exactly that radiation that had been picked up by the Penzias and Wilson antenna, which was aiming at something else! Precise measurements proved that the energy from the radiation corresponded to a temperature of 3 K, i.e. just above absolute zero. Like the heat from dying embers in a hearth, the microwave radiation testified to the existence of a blazing fire in the past.

That blazing fire and its remnants had, however, been predicted many years earlier. In the late 1940s, a Russian émigré in the USA named George Gamow was investigating the initial stages of the expanding universe and the conditions then prevailing. He concluded that if the universe had begun from a dense, superheated state, some associated radiation must still be found everywhere. Indeed, his calculations showed that its temperature should be just above absolute zero.

In addition to being a physicist of wide-ranging interests, Gamow was a great popularizer of science. For almost three decades, his bold prediction on cosmic radiation lay buried in the vast bibliography of theoretical physics. Just as in art and literature, it often takes time and a stroke of luck for great ideas in science to shine out. It is thus typical that the existence of cosmic microwave background radiation had been predicted twice, and yet was discovered by chance! In any case, the 1978 Nobel Prize for Physics was awarded to Penzias and Wilson, as well as to Pyotr Kapitas, a prominent low-temperature specialist from Russia. As often happens in life, luck rather than bold, pioneering thought yet again won the day.

But how does microwave background radiation reach as far as us? It began its journey right after the Big Bang. The early universe resembled a boiling sea, with light as its main ingredient. According to science, the phrase "in the beginning there was the light" does not seem far from the truth. The same sense is encountered in the poetry of Greek Nobel Laureate Odysseas Elytis:

In the beginning the light
And the first hour
when lips still in clay
try out the things of the world[5]

Light constantly interacted with the various matter particles such as electrons, quarks and others that no longer exist. Light radiation could not move freely due to collisions, and the universe was opaque.

In that critical period the number of photons rapidly stabilized, vastly outnumbering matter particles, so that for every proton or neutron there were a billion photons! That meant that the energy of light radiation prevailed over that contained in matter: it was a time of 'photocracy', when light reigned supreme.

As the universe expanded, however, the photons constantly cooled and lost energy. At some point, although the ratio to matter particles remained overwhelmingly in their favour—still by a billion times!—the stable energy of the latter prevailed. From domination by radiation, the universe came to be dominated by matter. In terms familiar to us from the evolution of modern society, this was the transition from the rule of light to the rule of matter, a transition which also marked the inception of the 'historical' universe.

We should note that the passing of the universe from prehistory into history is no mere figure of speech; it was then, approximately 400,000 years after the Big Bang, that the first neutral atoms were formed. So collisions between photons and matter became ever rarer. Photons ceased to stagger drunkenly around, travelling freely from then on. The universe became transparent. This period is quite rightly termed historical, since it is from then on that we can make direct observations.

Although we all know that the demand for transparency in politics is vague and somewhat hypocritical, in the history of the universe it has a specific meaning. From the moment light energy became transparent—a 'moment' that may have lasted centuries—it diffused everywhere, filling the space of a constantly expanding universe. That is the radiation that reaches us today, though with one essential difference: as the universe expands, light radiation constantly cools and its wavelength increases. So while the initial photons were in the region of visible light, they have now been converted into radio waves. They have become the paradoxical, impressive echo of creation.

Of course, tracking cosmic radiation and measuring its attributes have not been limited to the Penzias-Wilson horn antenna. November 1989 saw the launch of a satellite mainly charged with carrying out a detailed investigation. It went by the eloquent acronym COBE—Cosmic Background Explorer— and the sensitive instruments on board confirmed that cosmic radiation really does come from every direction in the sky at a steady intensity. This time its temperature was very precisely measured, coming in at 2.735 degrees above absolute zero. Indeed, the radiation's spectrum has what physics calls "black body" features, thus underlining its thermal origin. The maths shows that at the time light was decoupled from matter, the universe was roughly a thousand times hotter than it is today. In other words, data from the satellite left no doubt as to the nature of the radiation or its cosmological origin.

Nevertheless, one further COBE discovery shook both the scientific community and the general public when it was announced in 1992, to the point where many believed the golden age of cosmology had dawned. George Smoot, the University of California professor at the head of the research team that made the discovery, went even further. As he was to comment: "If you're religious, it's like seeing the face of God."[6]

And while the face of God does not appear moved either by human discoveries or sufferings, the COBE discovery really did concern a crucial moment in the evolution of the universe: the formation of the galaxies. That is, the appearance of certain early concentrations of matter, which constantly grew and were reinforced by gravitational attraction. As microwave radiation encountered concentrations of matter—or no matter whatsoever—on its journey, it was only natural for its intensity to vary. These variations resembled impalpable ripples on a boundless, uniform ocean, corresponding to temperature differences of no more than one ten thousandth of a degree.

These minuscule differences were what COBE managed to detect. Had it not been for the unrivalled precision of its instruments or the researchers' abilities, the discovery might have taken much longer, leaving a major gap in Big Bang theory.

7.4   Early Traces in the Formation of the Galaxies

The decisive significance of the COBE finds for cosmology and Big Bang Theory led to the need to analyse cosmic microwave background radiation in greater detail. To that end, two special space missions—NASA's Wilkinson Microwave Anisotropy Probe (WMAP) and the European Space Agency's Planck Mission—were placed in orbit in 2001 and 2009 respectively. With their special apparatus, they precisely measured the characteristics of microwave radiation and the anisotropy (directional dependence) displayed by it. From the data gathered they also determined the age of the universe, which came to approximately 13.8 billion years, while the WMAP space mission made an initial estimate of what the early, superhot and superdense universe consisted of: it included photons and neutrons, though dark matter was also heavily present. In any case, deciphering the will of God from the initial composition of the universe is far from easy!

But that isn't the impressive thing—it's that the temperature fluctuations recorded by detailed microwave radiation research made it possible to map the early universe. The resultant images (Fig. 7.2) exuded scientific learning and unprecedented beauty, inundating the planet and inspiring wonder and awe. It was no small matter: after centuries of hypotheses and theories, humans were able to see with their own eyes the specks on a paradoxical map and imagine the Big Bang that caused them. The mind-boggling thing was that by slow processes, billions of years later those specks would evolve into galaxy clusters and galaxies, stars and planets. And on one special planet, at some moment the miracle of life would burst into bloom.

That aside, the conclusions arising from the exhaustive study of cosmic radiation concerned more than how the galaxies were formed. The so-called wrinkles detected in the radiation's uniform distribution lent weight to a radical version of Big Bang theory known as Inflation Theory. According to this version of events, in the first hundredths of a second after coming into existence, the universe suddenly expanded at a break-neck speed much faster than that seen today. For all its infinitesimal duration, this "expansion within an expansion" solved many existing problems, but led to the worrying conclusion that our own universe might just be one among many others created like magic bubbles during the inflation phase. In other words, as if it weren't bad enough that the Earth had been ousted from the centre of the universe, and then our Solar System from the centre of the galaxy, and that our galaxy turned out to be one among billions, we might have to get used to the idea that our universe isn't even the only one in existence!

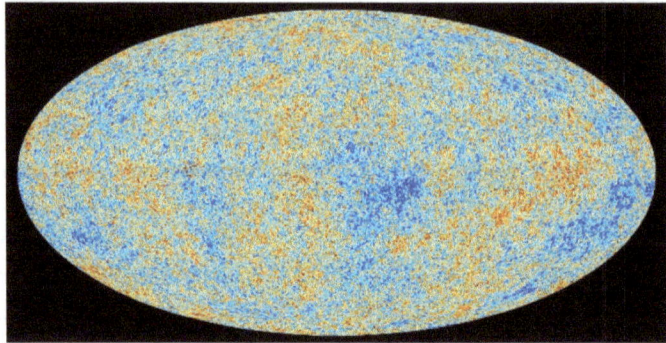

Fig. 7.2 Map of the cosmic microwave background radiation, imprinted on the sky when the universe was 370,000 years old, showing tiny temperature fluctuations that correspond to regions of slightly different densities. (ESA and the Planck Collaboration/NoirLab/CC 4.0)

The sensational fossil of light known as cosmic radiation has thus turned out to be a rich source of information and knowledge. How could it be otherwise, considering that the cosmos is revealed to us via light? Besides, the light fossil has a poignant attribute all of its own: while ordinary fossils are kept in special conditions and only ever touched by researchers, it surrounds us almost everywhere we are.

Even in a room at home, every cubic centimetre of air contains a few hundreds of the primordial photons that were true witnesses to creation! They have certainly grown weak on their unbroken journey over the past 14 billion years, though photons themselves don't age, of course; time does not pass for light. But just think of how much that primordial light could tell us! It has not only experienced the first moments of creation, but also the constant expansion of the universe. It has marvelled at the birth of stars and the violent collisions between galaxies, seen any number of comets and asteroids, and touched stunningly beautiful nebulae. Meanwhile, on a small, insignificant planet, it has felt that some curious, persistent stirrings would much later lead to the miracle to be called life.

7.5   The Language of the Stars

The stars and their light have long accompanied mankind's presence on Earth. They have been the source of myths and inspired artists and poets. Yet at the same time, skywatching has never ceased to have practical aspects to it. Ancient travellers and seafarers steered their course by the position of the stars; Christopher Columbus might never have reached America without their assistance. Astrology now takes advantage of people's insecurity and gullibility, but for a long time it served as a companion to astronomy, offering its own valuable observations to our knowledge of the skies.

All the same, understanding the nature of the stars took a great deal of dogged effort. As Antoine de Saint-Exupéry's Little Prince observes: "All men have the stars [...], but they are not the same things for different people. For some, who are travellers, the stars are guides. For others they are no more than little lights in the sky. For others, who are scholars, they are problems."[7]

Satisfactory solutions to those problems were found thanks to the information brought to Earth by an important messenger: the light of the stars. That alone speaks to us of the structure, the past or the future of the stars. And the same applies to our neighbouring planets. Light from them (by reflection, as they don't produce their own light) is our best source of information on Mars and Venus, Saturn and Jupiter. Since the distances light has to travel from them to the Earth are short, we don't learn of their "ancient" history, as with the stars, but of their very recent past, in the order of a few minutes ago. It goes without saying that effectively interpreting light signals involves several branches of physics and chemistry, including thermodynamics and spectroscopy, optics and quantum mechanics.

At any rate, it is no exaggeration to claim that the stars speak the language of light. Yet that language only became compatible with science after the invention of the telescope. The moment Galileo turned his spyglass on the heavens and discovered the hitherto unseen moons of Jupiter, a new world emerged. Telescopes have been improving ever since; new techniques and types are constantly being tried out, and observation capabilities now go beyond our wildest imagination.

One major discovery was to play the leading role in this evolution. As we have seen, visible light only accounts for a small part of the entire electromagnetic spectrum. The visible area is bounded by infrared radiation in the lower frequencies, and by ultraviolet higher up. On the other hand, the imposing totality known as the electromagnetic spectrum begins with much shorter frequencies, also including radio waves, ending at the opposite extremity in high energy and high frequency X and gamma rays. With its all its benefits and perils, contemporary high-tech culture revolves around the use of electromagnetic waves for everything from TV to mobile phones, and from diagnostic medicine to radar.

All kinds of electromagnetic waves perpetually criss-cross the entire universe. Even sunlight, which reveals the visible world to us, is but a small part of the solar spectrum. As Arthur C. Clarke stresses: "What we call 'sunlight' is only a narrow span of the entire solar spectrum — the immensely broad band of vibrations which the Sun, our nearest star, pours into space. [...] Think of one octave on the piano — less than the span of an average hand. Imagine that you were deaf to all notes outside this range; how much, then, could you appreciate of a full orchestral score when everything from contrabassoon to piccolo is going full blast?"[8]

So right from the outset, we have established that the language of the stars displays a far wider variety of gradations and concepts than one might suppose at first glance. It is as if science has helped us to move from the simple expressions of primitive humans, who only talked about mundane things, to the language of Homer and Shakespeare.

All the same, the language of the stars does have its own syntax: the laws of electromagnetism. Just like every other kind of electromagnetic radiation, visible light always travels at the same incredible speed of 300,000 km a second. In the time it takes to blink, a light wave can go round the Earth several times.

Modern instruments analyse the language of the stars in all its infinite variety, probing deep into the vastness of space to reveal an entire universe that only recently lay unseen. For all its enormous achievements, optical astronomy is now reminiscent of a drunkard looking for his lost keys directly under a streetlight, simply because that is where he can see best.

Photographic plates have also been of great assistance to astronomy, since they can be exposed to starlight for several minutes on end. Just as at the beginning of history, we have now moved from the speech of immediate observation to the invention of writing.

In fact, in modern telescopes photographic plates have given way to CCDs (charge-coupled devices)—a new type of sensor capable of gathering light continually, which has aptly been dubbed a digital camera obscura. With its silicon retina and high-tech computer circuits for nerves, this mighty eye can display screen images of stellar objects that send very little light to our noble planet. Using the right programmes, we can even edit an image and enhance its fainter areas. It has thus become possible to observe fine details on the surface of planets, or ejections of matter from distant galaxies.

The written language of the stars gained new capabilities when telescopes were combined with spectrographs, which analyse the white light from the Sun or stars into its constituent colours. The stars have characteristic colours that range across the spectrum depending on their surface temperature. A blue star is anything up to 20,000 K hot, while a red one is a mere 3,000 K. The Sun is somewhere in the middle, at a temperature of around 6,000 K. Spectroscopy has also been used to determine the chemical composition of the stars, which has been found to consist mainly of hydrogen and helium, with small admixtures of other elements.

The abundance of hydrogen, the simplest element in the universe, is evident from the fact that its constant hum (the so-called 'song of hydrogen') is picked up by terrestrial radio telescopes at one specific frequency. So apart from language, the world of the stars also has its own songs. Somewhat optimistically, in the belief that the song must be familiar to extraterrestrial civilisations, its frequency was even used as the basis for an exhaustive investigation of the signals they might be sending us. All in vain—the only positive point was that organic molecules were detected in space, once again with the aid of radio telescopes, thus lending weight to the view that the miracle of life is not exclusive to our world.

Recent decades have seen another radical development to add to the terrestrial telescopes scattered across our planet's mountaintops and plains. For all the improvements constantly made to them, the problem is that the Earth's atmosphere absorbs a large part of all incoming radiation. In other words, it cuts entire phrases and local dialects out of the language of the stars. But in the space age that obstacle has been overcome: there are now dozens of satellites and orbiting observatories acting as human eyes in or above the atmosphere. And being sensitive to infra-red or ultraviolet radiation, and even to X and gamma rays, they can record hitherto unseen activities out in the vastness of space. So starspeak has not just been enriched in terms of content, but also of communicative competence.

All the same, it's not worth our while dwelling on the ways in which the language of the stars—light and electromagnetic radiation—becomes scientific knowledge. Just as in literature, what counts is the narrative itself, the inner world and passions of the heroes, or the implied meaning behind words. There is no disputing the fact that the universe as described in the language of the stars and nebulae goes well beyond the bounds of the most gifted imagination. It often borders on the incredible, placing the age-old questions about its meaning or purpose on another footing. "Poets," notes Richard Feynman, "say science takes away from the beauty of the stars – mere globs of gas atoms. Nothing is 'mere'. I too can see the stars on a desert night, and feel them. But do I see less or more? The vastness of the heavens stretches my imagination – stuck on this carousel my little eye can catch one-million-year-old light. […] Or see them with the greater eye of Palomar, rushing all apart from one common starting point when they were all perhaps together."[9]

Yet even if the poets remain indifferent, the language of the stars—which is their light—is writing a breathtaking novel. Its pages are constantly being decoded, but are seemingly without end, and new ones are forever coming to light. As for the author, sadly neither her intentions nor her identity are a matter for science.

7.6   The Past Surrounds Us

Light and the various forms of radiation reaching terrestrial and space tele-scopes have led to deeper knowledge of the starry sky. Yet the sensational thing is that while light is our only source of information on the stars, it never depicts their present condition. It always depicts their past. This devilish game of light rests on the fact that while its speed is unimaginably fast, it is not infi-nite; and since the distances from the stars and galaxies are enormous, their light takes a long time to reach us. It follows that we see any given star, galaxy or quasar as it was when light left its surface. In essence, our eyes or astro-nomical instruments record the past of stellar bodies. As for their present condition (in our terms), we can only hypothesise.

It is a triumph of the human mind that these hypotheses usually turn out to be correct. Thus the supernova that suddenly flared up in 1987 in the Large Magellanic Cloud (Fig. 7.3), 160,000 light years away from us, confirmed prevailing theories on stellar evolution, even though that cataclysmic event had actually taken place 160,000 years before, when human beings were still learning how to use fire. To take another example, the red giant called Antares in the constellation of Scorpio is even visible to the naked eye. Since it is a few hundred light years from Earth, the reddish light now reaching our eyes or telescopes set out on its long journey when Constantinople was still the capital of the Byzantine Empire. So although stargazing is a great joy, it also is somehow unnerving. When we marvel at the stars in the night sky, we are touched by light that may have set out on its journey when the pyramids were being built, at the time the wheel was invented or when Plato was teaching in ancient Athens.

As for our nearest galaxy, Andromeda, we can turn the argument on its head: if it is inhabited and technologically advanced, then its state-of-the-art telescopes will be capturing the image of planet Earth as it was two million years ago. They won't find great geological differences. But human beings had only just stood upright and begun to walk on two legs. Only if they have evolved in a way similar to us will the aliens in Andromeda be able to imagine that humans have since come to dominate the world, conquered the air and discovered the fundamental laws of the universe.

So one paradoxical consequence of the colossal yet finite speed of light is that we "live" in the universe's past, but never in its present. We even see our neighbour the Sun at a delay of 8 min, which is how long it takes its light to reach the Earth. Some of the stars we wonder at on clear nights may long since have exploded, or quietly disappeared into black holes.

Fig. 7.3 Hubble Space Telescope image showing Supernova 1987A within the Large Magellanic Cloud, a neighboring galaxy to our Milky Way (NASA, ESA, Robert P. Kirshner (CfA, Moore Foundation), Max Mutchler (STScI), Roberto Avila (STScI), Copyright free)

At any rate, the fact is that modern radio telescopes can even track the furthest reaches of the universe. In other words, they can see several billion years back in time, to when the galaxies began to take shape. The photographs and treasure trove of information that the Hubble space telescope has sent back to eagerly awaiting scientists is just one impressive example. As we have already seen, Hubble has been orbiting 600 km above the Earth's atmosphere since 1990. Its advanced technological capabilities have yielded uniquely sharp images of the orgiastic processes going on in the vastness of space. What is more, it has managed to detect stellar objects ten times further away than had been previously been possible. We should also note that its 2.5 m primary mirror is the smoothest one of its kind ever built. Its resolution is so high that it can make out a firefly at distances in the thousands of kilometres!

Using Hubble's eyes we have seen galaxies up to 1.3 billion light years away, getting remarkably close to the early stages in the creation of the universe.

We obviously cannot say for certain how a distant quasar or ghostly nebula located by the telescope has evolved since then. But we do have millions of snapshots of the universe's past, and of various stages in the life of its stellar formations, enabling us to piece together a fuller picture of how it evolved. Due to its successful track record, Hubble's operation was greatly extended. It even continued to send impressive photographs back from the vastness of space when its "replacement" was already in the skies. Quite literally. In December 2021, the James Webb telescope was launched, rapidly reaching so-called Lagrange Point Two, 1.5 million kilometres from the Earth.

Point Two is an equilibrium point, where gravitational attraction from the Earth, the sun and the moon cancel each other out. The James Webb Telescope—regardless of its unfortunate name—is an incredible feat of technology with unique capabilities. Essentially speaking, it is not so much a replacement for Hubble as its highly worthy successor. Made up of eighteen gold-plated hexagonal segments, its mirror is 6.5 m in diameter, whereas we saw that the one on the Hubble was no more than 2.5 m (Fig. 7.4). The James Webb also has an ingeniously opening sunshield, which keeps its instruments and mirror at very low temperatures. Its most important feature, however, are its exceptional resolution capabilities, extending from the orange or red part of the visible spectrum to infrared. It is thus able to depict events and galaxies very far out in the universe, only a few decades after its explosive birth. The James Webb's goals include understanding how the stars and planets formed, and gathering extensive information on the characteristics of exoplanets and their ability to host life.

A few months after the James Webb was launched, in the presence of US President Joe Biden, NASA released the first photographs of an ancient cluster of galaxies in the depths of the universe. The enthusiasm and awe felt all over the globe was unprecedented. "I've spent fifty years on this topic [the universe]," said Dionysios Simopoulos, former decades-long director of the splendid Eugenides Foundation Planetarium in Athens, "but I've never seen anything like it. We're now talking about the birth, evolution and history of the universe. [...] These photos aren't just photos, there's a huge amount of background information in them."[10]

Recent years have seen China make a spectacular entrance into the field of astrophysics, alongside India's successful spacecraft missions, first to the moon in August 2023, and the very next month to the area around the sun. This has led the global community to the understandable impression that investigation into the wonders of the Universe is increasingly being restricted to the powerful states on our planet.

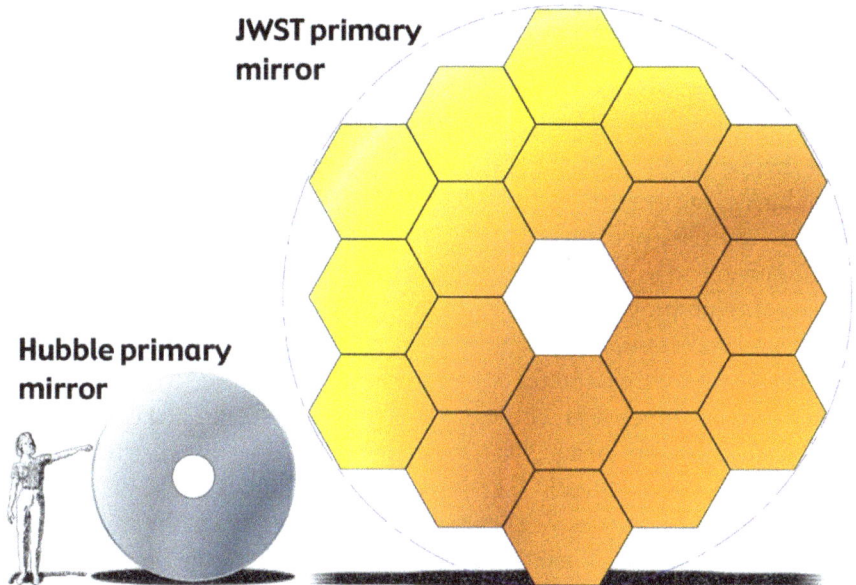

Fig. 7.4 Hubble-Webb mirror comparison: the collecting area of Webb's primary mirror is over 25 m squared, compared to just over 4.5 m for the Hubble. A larger mirror can collect more light, and thus detect dimmer or more distant objects (NASA-STScI/Copyright free)

Although this insight is not far from the truth, there is another side to it. Particularly over recent decades, the stunning discoveries in astrophysics and the enthusiasm generated by them have prompted nations of lesser economic or demographic clout to participate in related developments, either by establishing observatories or via international partnerships.

One characteristic example of this is the University of Crete. A few years after its foundation, in collaboration with the Max Planck Institute of Extraterrestrial Physics in Germany, the university acquired its own observatory atop Skinakas peak on Psiloritis (Mt. Ida), at an altitude of 1,750 m. The small telescope back then had its first 'cosmic' encounter with Halley's comet, whose orbit approaches Earth every 76 years. Since that time, thanks to constant upgrades to its optical capacity, the exceptional atmospheric clarity, and collaboration with the Greek Foundation for Research and Technology, Skinakas Observatory has made a substantial contribution to international astrophysics research. At the same time, it has recorded important locations and entities in the vast Universe at extremely high photographic resolution.

What is more, Skinakas Observatory boasts one unique feature that I think is worth mentioning. A few kilometres away, archaeological research has located an impressive cave known as the Idaean Antron, hiding a wealth of finds within it. According to Greek mythology, the cave was the birthplace of cloud-gathering Zeus, God of the Heavens. Many of Zeus's feats that were closely related to celestial phenomena—such as the ability to hurl thunderbolts—are described in Homer's iconic Iliad. Today, many centuries on, men rather than gods show the same reverence and curiosity towards the heavens, and Skinakas Observatory is a shining example of that tradition. In fact, in the near future its operation is to be substantially upgraded with the addition of cutting-edge scientific instruments such as polarimeters, which may even enable recording of the Big Bang in the faint light of Cosmic Microwave Background radiation. Furthermore, next to the first telescopes on Skinakas that have offered Greek and foreign astronomers the chance to make tangible contributions to modern discoveries, a new array of small telescopes is under construction. Despite being dwarfed by the giant ones on the high mountaintops of Hawaii and Chile (Fig. 7.5), their capabilities are impressive: in a matter of seconds, they can delineate an enormous area of the sky, equal in area to 400 full moons! Their purpose is thus to pinpoint the precise position of cosmic explosions, such as those caused when two neutron stars collide, emitting both light and gravitational waves. This combinational analysis of the light and space–time ripples caused by gravitational waves will open up new horizons in understanding the universe.

So if Zeus is still in the heavens, he will be pleased with developments that echo his own interests as God. Most of all, however, these developments meet the expectations of the demanding modern goddess known as science. On her horizons are many gifted mortals, expanding both our knowledge and the truth.

Fig. 7.5 The Milky Way forming an arc high above the antennas of the Atacama Large Millimeter/submilimeter Array operated by the ESO and partners in Chile. The arc is caused by the panoramic view of the image (E. Duro/ESO/CC 4.0)

In a time when praise is heaped on the resolving powers of the tele-scope, it's interesting and amusing to reflect what its early prospects were. "I have made a telescope," wrote Galileo four centuries ago, "a thing for every maritime and terrestrial affair and an undertaking of inestimable worth. One is able to discover enemy sails and enemy fleets at a greater distance than is customary, so that we can discover him (the enemy) two hours or more before he discovers us, and by distinguishing the number and quality of the vessels judge of his force whether to set out to chase him, or to fight, or to run away."[11]

The fact that light is not diffused instantaneously enables the universe to keep its present face a jealously guarded secret. On the other hand, as if guilty over its stance, light allows us to see deeper into space and time. The starry sky is like a peculiar time machine. The deeper we humans penetrate space, capturing radiation from distant galaxies and stars, the richer the universe's past and structure appear before our dazzled eyes. And conversely, our own past is eternally travelling into the vastness of space. Indeed, the fact that electromagnetic waves carry music or human speech through the airwaves aids Leonard Shlain in his interesting musings: "Imagine the excitement that will be generated when some lone radio ham, on a distant planet orbiting a different sun from ours, one night just happens to turn on and tune in to Earth. What a surprise will unfold, because it is all there—the entire history of the twentieth century as well as music since the Renaissance. […]"

"Anyone receiving our early broadcasts would be tuned to musical trends and historical events that have already happened here on Earth. Because of the time it takes light to traverse space, they will not know the outcome; having to wait in nail-biting suspense, like children at a Saturday matinee, to find out who wins World War II or the answer to the crucial question of whether we ultimately destroy ourselves in an environmental apocalypse."[12]

At any rate, research in modern cosmology has turned physicists into a curious breed of archaeologists. They are the archaeologists of the universe. And just as classical archaeologists try and piece together an entire building and its era from a few finds, so modern physicists strive to penetrate deeper into the universe's past and describe its evolution on the basis of their own observations and theories. It's just that classical archaeologists never come directly face to face with the past, whereas the archaeologists of the universe are destined to be surrounded by it, to the exclusion of all else.

As we have seen, unearthing the past has now reached as far back as the universe's infancy, a mere few hundred thousand years after the Big Bang. The COBE satellite mentioned above has recorded slight ripples or "wrinkles" in the isotropic structure of cosmic microwave radiation, echoing the slow but steady formation of the galaxies.

Primordial light has thus revealed the fingerprints of the eternally Absent One. Human beings are forbidden from seeing Her actual Face. That is a boundary to the distant past once again set by light; before the galaxies were formed, light was trapped in its constant collisions with matter. The Big Beginning may remain accessible to the mind, but not to the senses or to our technological accomplishments.

7.7   Nuclear Reactors in the Stars

Down the centuries, the light of the stars and its origin have constantly preoccupied human thought. Yet it is only in recent decades, since the chemical composition and processes in the interior of stars have come to light, that the answers to our questions have moved from the realm of supposition to scientific certainty. Earlier views which attributed starlight to burning carbon or a magnetic field now sound naïve. That is hardly rare in the history of scientific ideas: as knowledge moves forward, earlier explanations—of heredity, for instance, or how earthquakes arise—appear anything from downright bizarre to quaint. And the reason for that is that we rarely take the trouble to view a scientific theory or even a historical event in the context of its own time. In fact, it's worth wondering just how many of our present beliefs will hold true a few decades or centuries from now.

Accounting for the light of stars was unarguably a triumph of the human mind, if only because the stars really are extremely far away. So far, our own solar system has been the only thing accessible to spaceships from Earth, which have had the chance to make valuable observations close up. Note that the Sun is 150 million kilometres from the Earth, or in astronomers' parlance a little more than eight light minutes away. But the next nearest star, Alpha Centauri, lies at a distance of 4 light years, or 250,000 times further away than the Sun. It would take the fastest spaceship 60,000 years to get there! But as our imaginary journey lengthens, things get tougher still: Sirius, the brightest star in the night sky, is almost nine light years from Earth. And the distances to the other stars in our galaxy run into the hundreds or thousands of light years. As for the far-off galaxies, they are up to billions of light years away from our own.

So the stars and the origin of their light are not immediately accessible to scientific investigation. The same will still apply even if the current speed of spaceships increases ten times over. In fact, our enormous distance from the stars is what makes them appear colourless, shapeless and one dimensional to the naked eye, like mere bright specks in the heavens.

Yet incredible as it may seem, those bright specks are entire nuclear reactors! Science now has no doubt about it: both simple and complex nuclear reactions occur in the interior of the stars, and it is they that are responsible for the enormous amounts of energy and light released. As a matter of fact, the story goes that the prominent nuclear physicist Hans Bethe was taking a stroll with his sweetheart the evening after he came up with this theory. "Look at how the stars are shining," the girl said in time-honoured romantic mood. "Yes, and I'm quite possibly the only person who knows why," he boasted. There are, of course, serious reservations as to whether Bethe's companion ever fully appreciated his self-centred response.

In any case, our own life-giving Sun stands as a good example of nuclear reactions. Like other stars it consists largely of hydrogen, the simplest element in the periodic table, with a single proton in its nucleus. But while hydrogen protons should repel, the high temperatures (of up to twenty million degrees!) and pressures in the Sun's interior force them to rub up against each other constantly and merge. These loving embraces between protons result in the highly significant effect known as fusion. In a succession of steps, the hydrogen cycle ends up creating helium nuclei, each of which consists of two protons and two neutrons. Vast quantities of energy are released at the same time.

But what is responsible for energy being released in the course of nuclear reactions? Quite simply, it is the loss of mass seen during these complex processes, which just leaves its equivalent behind in the form of energy. We have already seen how this is predicted by the general theory of relativity. It doesn't only concern nuclear reactions; it is a general feature of our cosmos, underscoring the fact that matter and energy are two sides of the same coin. Note that humans on Earth have yet to succeed in effectively mimicking the nuclear fusion in the stars, which produces incredible quantities of energy. Yet just one gramme of matter converted into energy generates a power yield equivalent to 250,000 tons of gasoline. That's all it takes to appreciate the amount of energy produced effortlessly and at no cost in the Sun, where 5 million tons of hydrogen matter are converted into the equivalent of 400 billion trillion kilowatt hours every second! And since that particular magnitude goes well beyond the bounds of normal human imagination, here's an instructive analogy: if there was a hyper-cosmic electricity company, it would be charging us an amount equal to several thousand trillion annual budgets for every second of the Sun's energy!

Luckily enough the Sun is fairly large, and will only be paying for its self-sacrifice billions of years from now, when its nuclear fuel runs out.

It goes without saying that not all of the energy released by the Sun reaches Earth. Only a few billionths of that enormous amount ever end up on the surface of our planet. Otherwise the planet would not exist, or at least not in its present form. Yet even that minimal proportion is ten thousand times more than the total energy expended by humanity on its actual or invented needs. So it's easy to account for the strategy adopted by many of the developed countries on the planet, often bordering on the obsessive, to convert solar energy into electricity and thus cover part of our energy needs. The conversion work is done by so-called photovoltaic systems that are constantly improving in performance, and already cover large expanses of ground and even rooftops. For the umpteenth time, this goes to show that light and its energy are bound up with everyday life and human progress.

As far as the energy released is concerned, the Sun is no exception in the skies. All of the stars are born of giant nebulae, which begin to contract under gravity. When the pressure of contraction puts the core temperature up to several million degrees, slumbering nuclear reactions are aroused. At some point the stars "turn on" their light, and then continue their lives and shining undisturbed, as the nuclear processes in their interiors constantly convert matter into energy.

Like the Sun, most of the stars in the universe belong to what is known as the main sequence. What characterises them is a state of equilibrium, so to speak, in which the constant inward pressure of gravitational collapse is offset by the enormous energy being released in their interiors. The surface temperature of main sequence stars can be as high as 30,000 K. Emitted light is temperature dependent, ranging from red to bluish. Sirius, Vega, Polaris and most of the stars we wonder at in the night sky belong to the main sequence, and may well live on brilliantly in dynamic equilibrium for billions of years. We should note in passing that there is also a large number of stars smaller and colder than the Sun, emitting much fainter light.

In any case, the stage in a star's life when it shines brightly and steadily does not last forever. At some point the raw hydrogen matter runs out, forcing the star to come up with new ways of surviving. In most cases nuclear fusion processes carry on, increasing in complexity: instead of hydrogen, which has run out, helium then begins to burn. New elements are produced in the star's interior, starting with invaluable carbon, but this time energy is produced at a much faster rate. The star feels an onrush of panic, as the spectre of gravitational collapse poses an ever-greater threat to its existence. The pressure in the interior not only counteracts gravity, but triggers massive expansion.

Energy is radiated from its surface at greater wavelengths, and the star assumes a reddish hue. It is now a red giant, one hundred times larger than it originally was. One well-known red giant is Antares, 600 light years away from us, which is a few hundred times bigger than the Sun.

The moment the thermonuclear burning of helium runs out, the star begins to burn carbon, i.e. the "ash" it has formed. Ever more complex elements make their successive appearance and go on to be burned up, such as oxygen and heavy metals including iron and copper.

Nuclear processes are thus not simply the root cause of stellar heat and light. They also end up synthesizing complex nuclei. This itself is an important step in the organisation of matter, a step which the Big Bang was unable to complete: conditions then only allowed for the formation of helium. Yet the need to create complex nuclei does appear to be extremely strong, and is satisfied in a complex manner in the interior of stars. It is a need related to the existence of human life, or so at least one bold, egoistic theory claims. According to the controversial "anthropic principle", the main object of the universe was the creation of intelligent life, and it is on that basis that all the process and characteristic values of physical constants should be interpreted. So the nuclear reactors in the stars produce the raw materials—more complex nuclei—that were at some point to prove vital for life to evolve, culminating in humanity. Incredible as it may seem, our existence was first moulded in the distant heart of the stars.

7.8   The Life and Times of the Stars

The light coming from the stars has revealed that an entire, wondrous world is active in their interior. Far from being inert, eternal creations, the stars live a turbulent life lasting millions or billions of years, mainly determined by nuclear reactions. And while larger stars are brighter, they burn up their fuel much faster. For instance, a star twice the mass of the Sun would be twenty times as bright, but would consume its hydrogen twenty times faster, and would have a lifespan of no more than 500 million years.

That fact places serious limitations on the planets where some form of life could evolve. If they are orbiting massive stars, the light will not last long enough for the evolution of life to go through its lengthy paces. The presence of life on our own planet—which is, for the time being, the only certainty—is due to the fact that the caresses of the Sun's radiation have lasted for millions of years. Calculations actually indicate that in the Sun, nuclear reactions must have begun at least four billion years ago. And while

600 billion tons of hydrogen are converted into helium every second, the end is fortunately still a long way off. The Sun is in venerable middle-age, and is expected to live for another 5 billion years. Then, having burnt up the hydrogen in its centre, it will change into an enormous red supergiant, vindictively torching the poor planets that were its faithful companions. This is known as the *ekpyrosis* or conflagration, i.e. the end of the world in fire, which Ancient Greek philosophy had conceived of early on.

All stars have gone through the red giant stage or will do so. As the late Vasilis Xanthopoulos notes, "The period when stars become red giants is the time in their lives when they fool around, thoughtlessly expending energy. The only difference is that stars go through adolescence after maturity, shortly before the end of their lives."[13] Vasilis, an ingenious physicist and colleague of ours at the University of Crete, lost his own life to the bullets of a psychologically disturbed student while he was teaching.

Yet the red giant stage does not last long. For all its spectacular character, it is a dramatic omen of death; a death both inevitable and occasionally violent. Depending on the mass remaining in the final phase of its life, a star may end its days as a white dwarf or a neutron star. Both are stellar remains, but of entirely different consistency. The important thing is that sadly, as is true of humans, their light goes out together with their inner life.

A white dwarf is a dense, faint star of comparable mass to the Sun, but about as big as the Earth, so that a clod of its "soil" weighs ten tons. Further gravitational crushing of it is prevented by the pressure of the electrons coexistent with nuclei in its ionised matter. A white dwarf first shines very brightly due to its high temperature, but as the nuclear reactions with in it have long since ceased, its light constantly wanes to the point where it goes out. And that is the final, bitter destiny awaiting our own Sun: in the distant future it will roam the skies as a dark dwarf, cold and invisible. Whether anyone aware of its enormous contribution will be around then we do not know.

White dwarfs are not the only end to the constant struggle stars wage against gravity. To avoid gravitational collapse, a dying star often finds itself forced to eject a large part of its matter out into space. If the mass left over from this stellar explosion exceeds a critical limit—roughly one and a half times that of the Sun—gravitational forces cannot be restrained by electron pressure, as in a white dwarf. Relentless crushing then reduces matter into an unusual state. The atomic nuclei begin to rub together, and interaction with the compressed electrons converts protons into neutrons.

This is how a so-called neutron star is formed. Its characteristic feature is its enormous density; though only a few kilometres in radius, it is millions of times heavier than the Earth. Because neutron stars revolve at very high speed and have a strong magnetic field, from time to time they "light up" the Earth with a beam of radio waves, and so are often called pulsars. They resemble lighthouses in space, whose revolving light may be trying to tell of their brilliant, adventurous past.

So a neutron star's presence in space is once again betrayed by its light, though this time in the form of radio waves. In fact, when the first such star was discovered in 1967, the regularity of its pulses briefly led to the hope that they were messages from an extraterrestrial civilisation. Until more considered thought held sway, the source of the radio pulses was even dubbed LGM, for "little green men". But to great disappointment, it was soon established that the pulses were being emitted by the melancholy remnants of an enormous star that had already ended its days. Since then, over three thousand pulsating sources of radio waves have been located in our galaxy, some of which are over a billion years old. One pulsar of especial interest has been detected in the Crab Nebula. It is the obvious remnant of the cataclysmic explosion recorded by Chinese astronomers a millennium ago. It spins on its axis thirty times per second and, as if desperate to make its presence felt, also emits flashes of visible light.

It follows that the stars are neither eternal, nor were they all born together in some distant past. Like humans they have their own life cycles, and their own places and dates of birth. In the phantasmagorical Orion Nebula (Fig. 7.6), the birth of stars is to this day a constant, majestic process. With every passing second, over 60,000 stars are born into the universe.

That being said, it is once again worth underlining that light is the only witness to the life cycle of the stars. It is light, first and foremost, that signals their birth, and light that is capable of giving us information on their inner life, their future and their death. The light of the stars, its intensity and brilliance, its characteristic frequencies and colours—all are linked not only to a star's size, but to its internal temperature, its distance and its stage in life. Light speaks to us of whether a star is in the bloom of youth or has already turned into a red-tinted giant; whether its brilliance is dying down because it has spent its nuclear fuel, or whether some higher destiny has ordained that its stellar matter be expelled in a blinding flash. Light truly is the rich, expressive language of the stars. The fact that that language speaks solely of the past does little to detract from its significance or grace. Many of the world's literary masterpieces refer to stories from the past. Yet they enable us to understand the present and predict the future, in as far as that is possible.

Fig. 7.6 Hubble Space Telescope image of the Orion Nebula (ESA-NASA/Copyright free)

7.9   The Stellar Apotheosis of Light

The conclusion of a star's life as a white dwarf, once it has spent its nuclear fuel, may be regarded as a calm and indubitably dignified end. But those that are at least ten times the mass of the Sun are threatened by the sinister, ruinous prospect of their matter itself collapsing. And in an effort to avoid that, they blast a large part of their stellar mass into space. This results in cataclysmic supernova explosions, known not only for the violent eruption of vast quantities of matter, but also for their extreme luminosity. Since they outshine the Sun a billion times over, they are easy to detect even in distant galaxies. One famous supernova was recorded by Chinese astrologers in the Crab Nebula in 1054 AD. In fact, the court astronomer paid with his life for the entirely understandable error of failing to warn the emperor of the star's sudden appearance.

Supernovae really do shine out unannounced in the celestial sphere. Where there was a common star or nothing at all, a "new" one makes its majestic appearance. Its brilliance lasts for a few months until, as often occurs in life, it lapses into insignificance once more. Yet it is an insidious insignificance: in place of a dignified neutron star, a supernova may leave its remnants behind in the form of a black hole in space.

But what leads a large star to this grandiose yet suicidal explosion? Over recent decades, dogged theoretical research coupled with the amazing growth of computers that enable us to perform complex calculations have led to a satisfactory understanding of the phenomenon.

As astrophysicists now believe, what leads a star to explode is the permanent inability of its core to escape gravitational collapse; following successive cycles of nuclear burning, the core comes to consist of iron, which cannot undergo fusion. While lighter elements continue to burn and equilibrium is maintained in the star's outer layers, the pressure exerted by the electrons and neutrons in the core will not suffice to restrain the colossal gravitational forces. Thus the star's core, which is no bigger than the Earth at this stage, begins to collapse and shatter. First to abandon the scene of the disaster is a flood of neutrinos. Generated by nuclear interactions, they carry vast quantities of energy to the outer crusts. This process accelerates gravitational collapse, which is completed in a few thousandths of a second.

However, as the density in the centre of the core constantly increases, even the penetrating neutrinos are incapable of escaping. At some point their trapped energy triggers a powerful shockwave, which moves outward at enormous speed. The incredible pressure it creates leads the star to explode so impressively. As the outer layers are violently blown apart, they gain dizzying speeds and form a cloud of hot gas that radiates light as it collides with interstellar matter. The stunningly beautiful nebula still billowing out in Cygnus (Fig. 7.7) derives from a 20,000-year-old supernova explosion. Meanwhile, at roughly 6 light years in diameter, the Crab Nebula (Fig. 7.8) is the luminous remnant (including a neutron star!) of the stellar explosion recorded by Chinese astronomers.

Although a few hundred supernova explosions have been recorded in other galaxies, they have been conspicuous by their absence from our own Milky Way. The last one was seen by Kepler in 1604—by an irony of fate, just a few years before the first telescope was made.

Another explosion that occurred recently in our galactic neighbourhood was thus quite justifiably hailed as a milestone in the history of astrophysics. For the first ever time, it could be recorded by telescopes and other experimental apparatus, and existing theories could be checked.

The explosion took place in the Large Magellanic Cloud, a mere 160,000 light years away from us, and was detected on the morning of 24th February 1987. Terrestrial and space telescopes of all kinds focused their attention on the shining dot. A few hours beforehand, sensitive probes actually built

Fig. 7.7 The Veil Nebula billowing out in Cygnus from a 20,000-year-old supernova explosion (Z. Levay, ESA/Hubble and NASA/Copyright free)

Fig. 7.8 A mosaic image of the Crab Nebula composed from exposures taken by the Hubble Space Telescope in 1999 and 2000 (NASA, ESA/Copyright free)

for other research purposes had already recorded the arrival of neutrinos, triumphantly verifying the fact that they are the first to abandon the scene of the cataclysmic explosion.

Now known as 1987A, the February supernova retained a fluctuating level of brightness until mid-June, when it calmly began to dim. Yet scientific excitement continued unabated for quite some time. This was the first generation that had had the good fortune to closely observe an explosion of this kind and garner such a wealth of information. The supernova's antecedent, i.e. the star that had exploded, had been recorded in the past and didn't take long to locate. It was twenty times the mass of the Sun, and the inconsequential life that held such a spectacular end in store had lasted roughly 10 million years. In that brilliant explosion—an apotheosis of light—the universe itself lent weight to the theories on stellar evolution, confirming the astounding capabilities of the human mind.

Although it seems as if exploding supernovae are acting wastefully when they eject vast quantities of matter into space, what they are doing is actually a fundamental process in the cycle of both the universe and life itself. And that is because the matter from the exploding star intermingles with primordial cosmic fluid, enriching it with traces of the heavier elements created by nuclear burning in the star's interior. In that way, previously inert space is converted into a vast chemical factory. Apart from hydrogen and helium, it then has the raw material necessary for the formation of heavier molecules such as water and ammonia, alcohol and solid salts.

It follows that any new second generation stars formed from the cosmic fluid will contain a variety of the heavier elements. The Sun itself is a second-generation star; and since the Earth detached itself from the same nebula that created the Sun, it is no wonder that elements with complex nuclei are present in its atmosphere, ground and oceans. Incredible as it sounds, the carbon in the paper this book is made of, the gold in our jewellery and the uranium in nuclear reactors all derive from stellar explosions in the distant, hazy past. "Mankind's dream of reaching the stars and touching them comes true every day, right down here on our own planet," writes Dionysis Simopoulos. "Pick a flower. Taste some fruit. Stroke your face. All of them are pieces of some star."[14]

The terrestrial biosphere thus owes its composition to large stars which, on dying, released the elements they fashioned with patience and foresight in their interiors. Their number included the essential ingredients that were at some point called upon to synthesize the culmination of matter: living organisms and human beings. However much it sounds like an enchanting fairytale, we are made of stardust. And it is to the death of some star that we owe our lives!

7.10   Large Stars and Their Dark Prospects

The light that is the language of the stars has readily revealed its wealth and power of expression, and helped humans to understand stellar life. Yet there are also mute astral bodies, devoid of speech, in whose domains light lies eternally imprisoned. And so it is: any matter remaining in massive stars after they have exploded as supernovae may be driven to collapse and, in effect, to non-existence, forming one of the notorious black holes that fill us humans with awe.

Although black holes are now a major object of theoretical research and a source of inspiration for science fiction films and novels, the idea that they might exist is much older. Even in the twentieth century black hole theory had a rough time, often coming close to sinking into oblivion before it became firmly established. This is far from rare in the history of science and ideas: human beings, their social institutions and often even the scientific community harbour strong pockets of prejudice and conservatism, which refuse to accept anything new.

At any rate, the first substantial calculations on the fate awaiting large stars were performed in around 1928, by a gifted young Indian physicist named Subrahmanyan Chandrasekhar. Impressively enough, he did the work as a young man, while en route from India to England to complete his studies at Cambridge. Chandrasekhar proved that the fate of a star in the final stage of life, once it has spent its nuclear fuel, depends on the remaining mass. He even calculated the critical limit determining whether it will end its days as a white dwarf or a neutron star. For large stars there is yet another, darker prospect: if their remaining mass is at least three times that of the sun, gravitational collapse cannot possibly be avoided, leading matter to cancel itself out. A "black hole" is thus formed in space, and a strong magnetic field is all that remains to indicate the star's disappearance.

This dark prospect for massive stars appears to have met with strong reactions from Arthur Eddington, an expert on relativity and star structure, as well as reservations from other contemporary physicists. Chandrasekhar was forced to change research field and was only vindicated much later, having already moved to the United States. In 1983 he was awarded the Nobel Prize, while the crucial limit beyond which a star is subject to collapse quite rightly bears his name.

Interestingly enough, black holes and their paradoxical behaviour are an inescapable consequence of the general theory of relativity. Yet since even Einstein himself viewed some of the solutions to his equations with suspicion, it took several decades for associated research to flourish and extend to actual calculations. It should be noted that as early as 1939, the consequences of a star more massive than the Chandrasekhar limit undergoing gravitational collapse had been described by the prominent American physicist Robert Oppenheimer. But the war followed, and Oppenheimer turned to studying atomic theory and heading up the bomb building project. Later, perhaps tormented by guilt, he became a noted peace activist. In any case, his fortunately entirely harmless papers on the evolution of stars came back into the limelight much later. In the 1960s, struck by Oppenheimer's work on the gravitational collapse of stars, another leading American physicist named John Wheeler turned his attention—and that of his students—to general relativity. He even coined the term "black hole" to render the paradoxical state of affairs whereby matter has undergone collapse. Some of his colleagues reacted against what they saw as possible sexual connotations, and tried in vain to replace it with the insipid term "hidden star". At any rate, a plethora of new astronomical observations and the discovery of quasars rapidly led to intense interest in black holes and the possibility of detecting them.

As subsequent research confirmed, after a supernova explosion the collapse of matter itself threatens the core of massive stars. This dramatic prospect is due to the fact that the star's lines of defence against gravity are weak, and nothing can restrain the successive contraction of matter. As it finally collapses into one single point ("the singularity"), what is formed is quite literally a black hole. As theoretical physicist Kip Thorne says: "Of all the conceptions of the human mind, from unicorns to gargoyles to the hydrogen bomb, the most fantastic, perhaps, is the black hole: a hole in space with a definite edge into which anything can fall and out of which nothing can escape; a hole with a gravitational force so strong that even light is caught and held in its grip; a hole that curves space and warps time."[15]

The black hole thus denotes an area in space of exceptionally small volume, where the colossal matter from a dying star cancels itself out. No visible or material information can escape its domain. For that to happen, the signal from the information would have to be travelling faster than the speed of light, and that is an insuperable obstacle. "They have built walls around me, high and thick,"[16] as C. P. Cavafy wrote. In this instance, the high walls are the handiwork of the gravitational field, which soars dizzily as one approaches the aptly named point of anomaly. At that point, which is of inconceivably minute dimensions, stellar matter lies forever imprisoned. Nor can it ever emerge or send information—as radio waves or light—to tell of its ordeal. In practical terms it has disappeared from the cosmos, leaving nothing but its strong gravitational field behind. Strikingly, if planet Earth were to end up in a black hole it would be no larger than an aspirin, but its gravitational field would be so strong that anyone standing half a metre away would be 100 billion times heavier than normal!

The boundary around the area in spacetime from which nothing can escape is known as the black hole's event horizon. It is not so much a physical surface as a kind of treacherous, unidirectional spacetime membrane. A bold astronaut could get inside a black hole by crossing the horizon, but would then be unable to escape. "Abandon all hope, ye who enter here" is the inscription at the gate of Hell, as Dante warns. In the astronaut's case, Hell is overwhelming gravitational attraction, which would be stronger at the feet than at the head. So before disintegrating entirely, her body would be stretched and subjected to dizzying forces of acceleration. Worst of all, her companions on board the spaceship would neither worry nor be able to appreciate her heroism, since they would continue to see her on the black hole's event horizon, motionless in space and time. Yet the astronaut just might turn out to be less unlucky than she originally thought: according to some theories, the entrance to a black hole might end in a small gateway leading to another point in the universe. Or to be more optimistic still, to an entirely different universe.

A black hole is thus a curious entity, isolated from the remaining universe. At its centre, time and space do not exist as absolute, distinct entities. Our knowledge and laws of physics are weak there; only the vainly sought-for quantum theory of gravity might yield some information on the goings-on within.

7.11 Photographing the Invisible

As mentioned above, the main feature of a black hole is that neither light nor any other form of radiation or particle is capable of escaping from its domain. Instead, anything crossing the event horizon is imprisoned forever. But if black holes are silent witnesses to the disappearance of large stars, how is it possible to detect their presence? Given that the language of light has fallen silent once and for all, searching for a black hole is like trying to find a black cat at midnight during a total lunar eclipse.

Luckily enough, a black hole is often revealed by the powerful gravitational field left behind after a star has disappeared, just as the Cheshire cat in *Alice in Wonderland* was betrayed by its grin. Detecting a potential black hole in space is thus achieved indirectly, via the strong and otherwise inexplicable gravitational effects seen in a given area. Another tell-tale sign is greed: when stellar matter is being devoured it heats up, sending out a would-be distress signal in the form of powerful X-rays. One undisputed black hole—21.2 times the mass of the Sun—has been located in Cygnus, well over seven thousand light years away from Earth. It is the invisible companion of a large star, whose motion is affected by its powerful magnetic field. As X-ray emissions reveal, stellar matter is spiralling defencelessly towards the black hole's event horizon. Near the centre of our galaxy there also appears to be an enormous black hole, possibly even more massive than four million suns; it is responsible for the intense emission of radio waves and infrared radiation seen in neighbouring areas.

Gigantic black holes may also lurk within the brilliant quasars. That is the only way to account for the enormous amounts of energy emitted by those enigmatic stellar bodies; for though they are much smaller than galaxies, they shine hundreds of times brighter. To take one instance, a black hole equivalent to at least six billion solar masses appears to dwell in the centre of quasar M87, which is a primordial galaxy. As observations from the Hubble space telescope show, the galaxy surrounds a gas disk 130 light years in diameter, rotating around a black hole. The only way to account for the enormous amounts of energy being released is if matter from the disk is constantly heading towards the hole.

In any event, for a long time the scientific community harboured considerable reservations about the actual rather than just the theoretical existence of black holes. It's characteristic that Stephen Hawking himself, who had contributed immensely to research into the topic, doubted that a black hole was responsible for X-ray emission phenomena in the constellation of Cygnus. In 1974 he even placed a handwritten bet with Kip Thorne, another leading astrophysicist, who believed precisely the opposite. Yet slowly but surely, developments in X-ray astronomy mainly due to a new high capability radio telescope vindicated Thorne: an invisible but very powerful black hole really did reside in Cygnus. The unthinkable was now an indisputable fact. Hawking himself was forced to admit defeat and even pay his bet: a year's subscription to Penthouse magazine!

That being said, the emergence of black holes into the forefront of science did not stop with Cygnus or Hawking's bet. Developments were incessant. So, in response to the growing interest in black holes, a large team of scientists achieved the inconceivable: they photographed the black holes that had already been pinpointed in the vast universe. Not their interior of course, which never lets even a single ray of light escape, but the shadow cast by a black hole on the luminous background of gases swirling around it.

Nevertheless, this accomplishment was far more difficult than it might seem. First of all, an international collaboration of scientists known as the Event Horizon Telescope (EHT) had to be set up. Under that name, numerous telescopes throughout the world were coordinated so that their observations gained the highest possible resolution and sensitivity. Using a technique known to specialists by the abstruse term radiointerferometry, EHT observations are of a resolution equal to that of a telescope as big as the Earth.

This virtual telescope initially turned its attention to the black hole which indications showed to exist in Galaxy M87, 55 million light years distant from Earth. The first photographs were released in 2019, much to the enthusiasm of scientists and the excitement of the general public. As large as our entire solar system, the black hole not only existed, but was casting its shadow on the shining disk surrounding it. As Vasiliki Pavlidou, professor of Astrophysics at the University of Crete, comments: "Turn on your computer and search for the image of the black hole at the centre of Messier 87 (Fig. 7.9), as recorded by the EHT. Take another look at the black smudge in the middle of the luminous ring and reflect on what you're seeing. Practically speaking, the dark, incomprehensible, terrible monster out there is now visible. And it's behaving just as outrageously as theory predicts."[17]

Here's the place for an impressive fact: a major contribution to the entire effort was made by Katie Bouman, a young graduate student in the Computer Science and Artificial Intelligence Laboratory at the famed MIT. It was she who created the vital algorithm needed to design the imaging methods, adding a further building block to the ever-increasing applications of Artificial Intelligence.

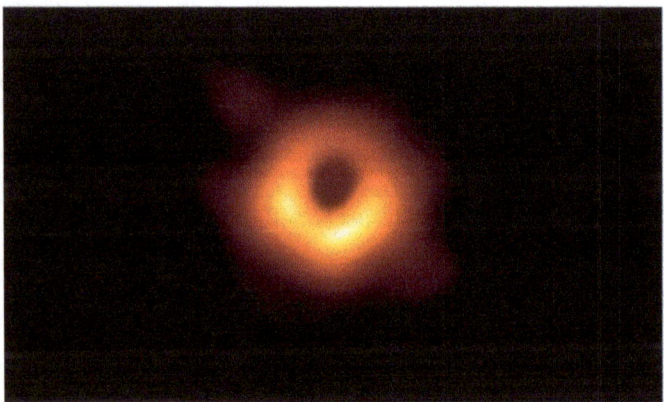

Fig. 7.9 Image of Sagittarius A*, the supermassive black hole at the centre of Messier 87, realised by synchronizing radio telescopes to create a virtual telescope the size of the Earth (EHT Collaboration/NoirLab/CC 4.0)

This may be the right moment for a digression, and one hard to keep in check, on the whole issue of Artificial Intelligence. While its positive applications are already apparent, particularly in medicine, its future frightens people more and more. And although these fears are not unfounded, it's worth noting that all the major scientific or technological advances of our time have been accompanied by similar concerns: nuclear energy, cloning and the leaps and bounds—for good or bad—in biology, the internet and its many side-effects. In all these cases the consequences of the advances were visible and positive, while vigilance and the steps society has taken or neglected to take so as to deal with potential dangers have been just as significant. Artificial Intelligence and its development should also proceed along the same general lines. In fact, I believe conditions are now ripe for the discussion on controlling AI to begin in societal institutions or global forums, before it's too late.

But let's get back to black holes and the mind-boggling photography of the invisible. The next step by the EHT was every bit as interesting, but also more difficult: an attempt was made to photograph the black hole which, by all indications, resided at the centre of our galaxy, some 27,000 light years from Earth. Known as Sagittarius A*, or Sgr A* for short, the difficulty of photographing it mainly lay in the presence of dust and stellar clouds that blurred the photograph like thick fog. As for the required resolution, it was like trying to take a picture of a tennis ball on the moon with a smartphone.

After painstaking processing, the photographs depicting Sagittarius A* were released in May 2022. As was expected, its structure was similar to that displayed by the supermassive black hole in M87, with its shadow once again visible in the fiery ring surrounding it. But one further conclusion arising from analysis of the photographs was of particular interest not just to scientists, but to the history of science overall: the mass, the size of the fiery ring and the size of the shadow were absolutely in line with what the General Theory of Relativity predicted. A century had already passed since Einstein's trail on planet Earth, but its imprint still retained every bit of its brilliance and uniqueness.

In rounding off the earth-shattering developments that black hole photography brought to the life scientific, it's worth mentioning a couple of young scientists, then both professors at the University of Arizona, who were pioneers in the entire undertaking. They are Dimitris Psaltis, a Greek, and his Turkish partner Feryal Özel. Shortly before presenting the photographs that depicted the black hole at the centre of our galaxy, they were asked in an interview whether there was any lesson for humanity from the amazing accomplishment. "I think the most important lesson doesn't have to do with astronomy, but with events on our Earth," Dimitris Psaltis spontaneously replied. "Science has progressed so far that it not only gives us material benefits, such as better communications or how to deal with the pandemic, but also answers to extremely important questions. As humanity, though," he went on, "we're at a stage where if we don't listen to science, we'll destroy our planet."[18] In our minds and actions, let both the powerful of the Earth and all of us ordinary citizens bear that stern warning in mind.

Anyone browsing through the scientific literature on black holes or the General Theory of Relativity will often come across the name of the British theoretical physicist Stephen Hawking. His case is truly unique and worth a special mention. Despite extreme physical handicap, Hawking's gifted brain remained constantly alert, and he never lost his passion for science or his sense of humour. He married twice, had three children and lived far longer than his doctors had predicted.

Hawking initially worked on the General Theory of Relativity and studied the mathematical anomaly synonymous with the Big Bang. Quite soon, however, black holes and their complex mathematical problems became the focus of his attention. In one of his key papers, Hawking proved that a black hole must emit radiation and particles just as a thermal body does, meaning that black holes are not that black! In fact, the smaller they are the hotter they get, and thus the brighter they shine.

All the same, readers shouldn't worry. This conclusion does not run counter to the definition of a black hole, nor to its incredible capacity even to trap light within its domain. It is simply a consequence of the oddities and omnipotent uncertainty principle gifted to the microcosm by quantum physics, which allows radiation to be generated from almost nothing. The emitted particles do not set out from the interior of the black hole, but from the empty space just outside its event horizon. So black holes always stay black!

I had the fortune to meet Hawking in person. In 1998, the eminent physicist was invited by the Department of Physics at the University of Crete to give a talk on the Theory of Everything. Difficult in many respects, the entire enterprise was made possible thanks to the late Thodoris Tomaras, then head of department, and proved to a historic moment for the university and the country. An unbelievably large crowd both inside and outside the lecture venue gave Hawking a standing ovation, welcoming him in awe and unbounded tenderness, spontaneously parting to allow him through in his wheelchair. That evening his large, expressive eyes were lit up with an otherworldly, almost mystical light and laughter. I felt the same smile later on, as he answered our questions or commented on matters of science and the world in his slow, metallic voice. His smile seemed to be reflecting the sublime smile of the universe which, in this case, coincided with the smile of humankind.

One thing regarded as indisputable is that black holes really are perpetual and possibly unique prisons of light. Being used to travelling and conquering the world, it is as if light wants to exact its revenge by making its eternal dungeon a place of the paradoxical and the fantastic.

7.12   A Dark Universe in Dark Times

It is not overstating things to say that recent years have been a dark period in human history. Violence and war keep breaking out in one corner of the globe or another, the nuclear arms race and associated threats show no signs of abating, while a deadly virus has sowed devastation and fear everywhere. At the same time, climate change has already shown the first signs of the dark future in store for us if we do not wise up.

Against that backdrop, readers will hardly be surprised by an unbelievable, stunning truth that has emerged from observing the stars and galaxies: the universe itself is "dark"! Put simply, it is dominated by unseen dark matter that is impossible for us perceive with our eyes or telescopes. To make matters worse, a kind of dark energy of unknown origin or nature is spread evenly through space, determining its future. Although the existence of dark matter has long been known to science, in recent years it has been corroborated by new observations and findings. It is encountered right across the universe, but is denser in the interior of galaxies. In any event, if we had magic glasses capable of penetrating dark matter, a galaxy very different from the one we now know would be revealed to our eyes. A dense halo of dark matter particles would appear to envelop the spiral disc familiar to us.

Digesting this new universal reality is far from simple. It means that the visible matter known to us—the billions of galaxies, planets and nebulae, comets, asteroids and interstellar dust—only correspond to a small proportion of the total mass in the universe, a proportion barely approaching five percent! Meanwhile, the terrifying 95% of the remainder is composed of mysterious dark matter and equally mysterious dark energy, along with its equivalent in mass. How unfair it is that the resplendent world of the stars and planets, lauded by our civilizations and our own souls, should be humiliated in this way by an insidious, dark reality!

Of course, one key question arises here. Given that dark matter is invisible to even our most powerful space telescopes, how can we be sure it exists? The answer is simple: because it exerts gravitational forces on the space around it, which cause anomalies in the motion of the stars and galaxies. It is reminiscent of the fictional invisible man, who makes his presence felt by moving furniture or nudging those in doubt. Thus, the velocity of a star orbiting around the centre of the galaxy—such as our Sun, for instance—cannot be accounted for simply in terms of the gravitational attraction of known, visible matter. Some additional force, obviously due to dark matter, appears to be acting. It even seems that galaxies could not possibly remain stable without the presence of invisible matter. Indeed, entire clusters of galaxies which should have broken up, remaining in love but worlds apart, display paradoxical stability thanks to the effect of dark matter. The leading French astrophysicist David Elbaz may not be wrong when he credits dark matter with a sublime mission: "Just as leaves could not exist without the branches and trunk of a tree," he writes, "we now know that no galaxy or star could exist without dark matter. The gravity of ordinary matter alone would not suffice to create the lumps in the universe if it had not benefitted from the gravitational wells of dark matter. So we owe a great debt of gratitude to this matter of unknown origin, since without it we would not exist."[19]

In this context it's worth making special mention of the constellation Coma Berenices, with which I have indissoluble bonds. Indeed, my first book bore the same name, was inspired by the constellation and went down well with readers. Well, the striking thing is that in the constellation Coma Berenices there is a cluster of galaxies numbering in the hundreds, but their mean velocity is so high that the cluster ought to have broken up long ago. It thus appears that some form of dark matter acts at sufficient intensity to hold the cluster together by gravity. All the same, its nature remains enigmatic. Perhaps those enigmatic characteristics can account for the curious attraction felt by the author for the ancient constellation, and for the form of Berenice herself. In fact, the author often thought that the woman accompanying him on his wanderings for centuries had something in common with Berenice, the star-spangled queen of passion and fate. In her too—as in Berenice—the past mingled indeterminately with the present, and here again was the melancholy of her eyes and the song dying in the silence of the world.

To get back to dark matter, the fact is that few astrophysicists now doubt its existence. Indeed, on the basis of evidence announced by the Planck space mission, it has been confirmed that dark matter accounts for almost 27 percent—a far from negligible proportion—of the entire mass in the universe. Yet despite some agreement that dark matter is a paradoxical yet actual entity, all manner of conjectures or theories have been propounded as to its identity. One scientific view, though not one widely accepted, holds that it consists of planets, supermassive black holes and dwarf stars. In other words, it does not greatly differ in composition from ordinary matter. Another view turns its attention to the ghost particles known as neutrinos, which abound in the universe and can pass unimpeded through planets and starts—as well as through us. As proved experimentally, however, neutrinos have infinitesimal mass, and taken together may contribute to the large quantities of dark matter apparently spread across the universe.

That being said, more hope for interpreting dark matter is held out by the existence of exotic, so far unknown particles. The theory hypothesizes that these exotic particles were abundant in the primordial cosmic soup following the Big Bang, i.e. the creation of the universe itself. The problem is that while these particles are persistently sought after in terrestrial accelerators, which create conditions similar to the Big Bang, they have yet to offer any sign of their presence. So they may well simply be scientists' wishful thinking. The despondence expressed by one of the pioneers in related research is typical: "'On Mondays, Wednesdays, and Fridays I think it's particles, and on Tuesdays, Thursdays, and Saturdays I think it's Jupiters.' How about Sundays? 'That's when I consider the possibility that there is no dark matter at all, and what we really don't understand is the laws of gravity.'"[20]

The last point was more than just a joke. One team of scientists sought an interpretation for dark matter in errors or weaknesses in the General Theory of Relativity that accounts for gravity. Wasted effort! Both the discovery of gravitational waves in 2015 and the recent photographic imaging of black holes, in absolute agreement with the Theory, put Einstein's ingenious insight on a shining, unshakeable pedestal once and for all.

7.13   A Dark Sea of Energy

Be that as it may, the sense of a "dark" universe is not simply limited to the existence of dark matter. Until recently, an incredible new surprise awaited both astrophysicists and public opinion. Named "dark energy", it poses a conundrum both in terms of its origin and its significance in understanding the universe. The fact is that observations collated over recent decades from powerful terrestrial or space telescopes have come to a startling conclusion: the expansion of the universe following its creation from the Big Bang is not slowing down over time, as gravitational attraction between galaxies would dictate, and as anticipated by scientists. Instead, it is accelerating! It appears that some paradoxical form of energy with repulsive properties overpowers gravitational attraction and forces the galaxies to retreat from each other at ever greater speed.

Once again, it was light that provided the indications as to this radical overturning of our convictions. In fact, in this particular case, it was the light from the stars. Analysis of it can determine both the distance at which a star lies and its recessional velocity due to expansion. But depending on the distance, the light now reaching Earth may have set out when the universe was younger. It is thus possible to compare the rate at which that same universe was expanding in different periods of its history.

Observations for this vital comparison focused on supernovae, whose brilliant explosions are often recorded in distant galaxies. They are like candles, suddenly lighting up in the countless galaxies to illuminate the universe's course through history. But while the scientific world was awaiting confirmation that the expansion of the universe caused by the Big Bang was constantly losing momentum, the observations showed the opposite. From some point onwards, the universe has been speeding up rather than continuing to slow down! The galaxies are retreating from each other at ever faster rates.

It should be noted that the first suspicions as to this incredible fact were expressed at the end of the previous century by two independent research teams: one was based in a state-of-the-art observatory on a mountain in Australia, while the other used a large telescope in Chile as its main source of information. Unable to believe their own findings, the researchers exhaustively checked every other potential interpretation or error in their calculations. But when you have eliminated the impossible, as Sherlock Holmes insisted, whatever remains, however improbable, must be the truth.

The truth, then, is that the expansion of the universe is accelerating, as if a powerful antigravity force is pushing the galaxies apart. A titanic struggle between this repelling force and the gravitational attraction know to us is being fought in the vast expanses of the universe, with the eventual winner still to be decided.

New finds, particularly from the Hubble space telescope, have confirmed that the expansion of the universe is accelerating. It has even been calculated that the critical period when the acceleration began occurred around five billion years ago. The strength of the dark energy causing the galaxies to retreat is also impressive—it appears to be close to 68% of all matter and energy in the universe. By way of comparison, let's remember that dark matter makes up 27%, while the visible matter familiar to us barely comes to 5%.

At any rate, it's worth stressing that suspicions as to the existence of some mysterious antigravity force go back much earlier. The fact that the universe could not be static, but was instead subject to constant expansion, could be deduced from the equations in general relativity theory. But that conclusion ran contrary to Einstein's sense of science. The ingenious physicist thus introduced an arbitrary constant into his equations, restoring equilibrium to the universe. Symbolised by the Greek capital letter lambda (Λ), this cosmological constant corresponded to a force opposing gravity, an "antigravity" of unknown origin.

A decade later, the triumphant discovery that the galaxies were receding— i.e. that the universe was expanding—led to the cosmological constant being abandoned, and to Einstein admitting that it had been the biggest blunder in his life. Nowadays, few still doubt the existence of some kind of cosmological constant or equivalent "dark" energy. Einstein's blunder could prove to be yet another sign of his scientific intuition, which was truly unique.

In any event, another interesting scientific view attributes dark energy to a different cause: the existence of a mysterious field answering to the wonderful name 'quintessence'. In Aristotelian philosophy the word referred to the aether, and was the fifth fundamental element in nature. If nothing else, it would come as a relief to feel that the universe—or our short lives, at least—were governed by some quintessence. The author had already experienced the quintessence of all the moments that she had granted, delighting in her inner song and the grace of her movements. And yet he, the author I mean, was often in doubt as to whether she was an actually existent woman, or the creation of some aetherial field.

There's no room for delusions. Over recent decades, research and observations by scientists have brought to light an incredible, mind-boggling truth: the universe we can see is only a small part of the actual one. Invisible dark matter and a kind of dark energy predominate throughout its vast expanses. It is as if the universe itself wants to be in step with our own dark times, which belittle and buckle our certainties and our dreams.

There is, however, one major difference. The structure and characteristics of the universe are beyond our influence; our only hope remains in better understanding them. Yet human life itself can to some extent gain a new meaning, as long as justice, respect for the environment, cultural values and the pursuit of peace prevail in humanity's journey. Perhaps then, many of the mysteries of the universe will remain unresolved. But humanity itself will gain more certain and, above all, brighter prospects.

7.14   Our Universal Prospects

Unlike the future of mankind, which depends on so many parameters as to be unknowable, the impressive thing about the history of the universe is that it depends on one magnitude alone: the current mean density of its matter. In fact, this also regulates its geometrical structure.

As we have seen, the universe had its beginnings in an inconceivable Big Bang around 14 billion years ago, and has been expanding and cooling ever since. Even today, the galaxies overcome the pull of gravity and are receding at high speed. For gravity to prevail at some point, and for the expansion to stop, the density of the universe will have to exceed a minimum "critical" value.

This critical density has been calculated theoretically, on the basis of the laws of gravity and the universe's current rate of expansion. It is an infinitesimal number. In normal units of density, i.e. grams per cubic centimetre, it can be shown as a fraction with a denominator of one followed by thirty zeroes. On average, this corresponds to around five atoms per cubic metre of the vastness of space.

Things are simple now—or at least at first glance. If the universe is denser than the critical value, then it is closed and finite. As it continues to expand, it is like the surface of a constantly inflating balloon.

Nevertheless, at some point the gravitational forces restraining the current expansion of the universe will prevail, and the events characterising its evolution will go into reverse: gravity will bring about accelerating contraction to ever smaller dimensions, and the universe will grow increasingly hotter and denser. Ever captive to its gravitational pull and in contrast to the Big Bang that created it, the universe will end in an equally indeterminate Big Crunch. Whether that will lead to a new Big Beginning and a universe in some way reminiscent of our own is too risky to predict. Not for any lack of arrogance in science, but because a quantum gravity theory capable of describing the state of matter in the extreme conditions of a Big Crunch has yet to be formulated.

The closed universe model—in never-ending cycles, what's more—has an indisputable aesthetic and philosophical beauty to it. Besides, a pulsating universe gets round the problem of its own birth, evoking a sense of eternity. But it does not appear to agree with the scientific data. The actual density of the universe is thus around 50 times less than the critical level that would lead to a reversal of expansion.

New hopes of bridging the gap have been offered by the presence of invisible dark matter, the reality of which is no longer disputed. And while its nature and origin remain unknown, it does of course exert gravitational forces. Yet even by the most optimistic estimates, the presence of dark matter barely raises the present density of the universe to about 30% of the critical value.

So for quite some time all the indications were that the universe was open-ended. Yet while the open universe's life will be infinite, it will be a life in name alone. The nuclear reactions in the stars will gradually burn out, leaving various kinds of ash behind them: neutron stars, black holes and some matter particles. Cosmic radiation will never disappear, but its energy will constantly wane.

In any case, the earlier debate on a closed or open universe has been stirred up by a recent discovery. As already mentioned, the expansion of the universe appears to be speeding up rather than slowing down. This apparently applies to the last 5 billion years, and is linked to the existence of dark energy, which serves to repel and outflank gravitational pull.

So a new possibility has emerged for the structure of the universe, and hence its future. It appears to be neither open nor closed, but flat! Its geometry is Euclidean. As it expands, it resembles an infinitely expanding, thinning rubber sheet. However, the case for a flat universe requires that its density be at the critical point, or at least very close to critical. The chances of satisfying that condition appeared negligible in the past, but the existence of dark energy and the mass corresponding to it makes up the difference necessary to equal critical density. So the universe has won its much-desired flatness. But why much-desired? Simply because two independent yet equally important paths in cosmology have led to the same need.

In the first place, the clearest picture of the early universe comes from measurements of the cosmic microwave background (CMB). As we have seen, this is the radiation left behind by the Big Bang, revealing the features of the universe when it was just 400,000 years old. Recent satellite measurements of CMB fluctuations have confirmed earlier indications, strongly leading to the conclusion that, in all probability, the geometry of the universe is flat!

One further line of argument seems to lead to an admirable consensus. And that is cosmological "inflation" theory, which is far more popular than the version more familiar in everyday housekeeping. A daring view that has solved many of the enigmas in cosmic evolution, it argues that immediately after the Big Bang, the universe underwent superfast expansion. This was an expansion within an expansion, rendering its geometry flat and regular. The gravity of dark matter had an important role to play in forming the structure of the cosmos.

Yet how did the density of the universe happen to be selected with such incredible accuracy that it coincided with the critical level? One answer—if it can be considered as such—is given by the renowned "anthropic principle". Only a universe like ours is capable of surviving long enough and containing the required matter for stars and planets to form. In other words, it satisfies the conditions necessary for the phenomenon of life to blossom on Earth, and for intelligent beings to evolve. With the arrogant hypothesis that the universe came about so that we could exist, our presence on Earth gains a whole new dimension. Though we may not have been worthy of the great honour in store for us, we puzzle over the origin of the universe and ask questions about its potential future. "The effort to understand the universe," notes Steven Weinberg, "is one of the very few things that lifts human life a little above the level of farce and gives it some of the grace of tragedy."[21]

Living in suspension, exiled from God and the centre of the universe, humans nonetheless dare to ask questions touching on their own existence and that of the Cosmos. The answer as to whether that is their grace or their tragedy goes far beyond science.

7.15 Mankind's Whimper and the End of Light

Recent discoveries appear to have sidelined the traditional dilemma over whether the universe is open or closed. It is flat. Besides, that is what the dominant theory of its evolution demands. To sum up the picture, the total mass of the visible and invisible universe falls far short of rendering its density critical. So at least for the time being, the case for a closed universe reverting to its beginning has to be ruled out.

All the same, the existence of repulsive dark energy puts the fate of the universe on a new footing. Right this moment, expansion is accelerating; and because the density of matter drops radically as space increases, dark energy may dramatically outflank gravitational attraction. The rate of expansion will then become explosive.

An ominous new prospect then opens up for the future of the universe. Rather than a Big Crunch, a Big Rip may lurk at its end. "And, behold, the veil of the temple was rent in twain from the top to the bottom," as the Gospel according to Matthew says. The tremendous effect of dark energy will at some point rip the galaxies apart, solar systems and their planets will dissolve, and all material entities—right down to atoms—will disintegrate into more elementary components: quarks, electrons and neutrinos.

Yet before the absolute end, an impressive phenomenon will indicate the ever-greater rate of expansion: one by one, the galaxies now seen by telescopes will disappear over the horizon. That's because their light will be unable to compete with the explosive expansion of space, which will lead star systems to recede faster and faster. Indeed, any astronauts still around may be able to relive the old illusion that our galaxy was the only one in existence. But even our own Milky Way will not escape the bitter cup: under the relentless pressure of dark energy, its own stars will at some point be forced into oblivion, and will cease to shine in the night sky. T. S. Eliot seems to have envisioned such a prospect for the universe, in the lines:

This is the way the world ends
    Not with a bang but a whimper.

Though the whimper at the end of the universe may be heard in several billions or even trillions of years, the whimper of mankind will almost certainly be heard much sooner. As has already been mentioned, the Sun is now in venerable middle age, with another 5 billion years left to live. Then, when it has burned up all of its hydrogen, it will turn into a bloated red giant. Its dimensions will grow so great that they will swallow the Earth and the other inner planets. Their fate will be a torching, an end of the world in fire. "Dies irae, dies illae / Solvet saeclum in favilla" ("The day of wrath, that day/Will dissolve the world in ashes"), as an apocalyptic medieval hymn says. Prior to this, perhaps in punishment for our folly, the Sun's temperature will slowly but steadily begin to climb; and rather than being life-giving and beneficial, its radiation will turn our planet into an inhospitable hothouse like Venus. Plants will wither, the oceans will boil and a thick layer of carbon dioxide will blanket the once unrivalled beauty of the Earth.

As a matter of fact, some imaginative scientists have pointed out ways that would enable to human race to survive the Sun's fiery vengeance. These include colonizing our colder neighbour Mars—which is in reality a pipe dream—as well as placing Earth in a safer orbit, to be achieved with the aid of rockets and an asteroid! It goes without saying that all of the above are reminiscent of daring pages from science fiction.

But even moving as far as we can from the Sun's scorching breath will do little to help our mental well-being. It is to the light of the Sun that we owe our existence, and from its light that we have learned of the world and seen its images. It will be tragic to see the star that has accompanied our every step in life from the very beginning assume the guise of a red giant, and then degenerate into a white dwarf no bigger than the Earth. Even that will grow dimmer over the centuries, before going out for good.

On the basis of observations and prevailing theories, then, we have roughly sketched out the future awaiting the universe. All the same, we hardly need stress the uncertainty surrounding such cosmological predictions. Even if dark energy is somehow shown to exist, we cannot rule out the possibility of it changing into an attractive force at some point in the future. The reversion to an infinitesimal universe in the Big Crunch will then be inescapable. Then again, nor will the journey to an open, infinite future necessarily take the path we have described. For instance, astrophysicist Fred Adams predicts that already dead matter will collapse into black holes. But after trillions upon trillions of years, the black holes will themselves disintegrate into stray particles; and these in turn will decay to leave a featureless, infinitely large Void.

It goes without saying that both the complexity and the unimaginably far reach of theories on the future of the universe will have readers scratching their heads. And while in some cases these theories do not lack the sheen of science, they are often reminiscent of novels in fantastical worlds with imaginary entities as heroes. Investigating the future of humans often also fails to avoid such extremes. As one "scientific" prediction claims, it could be that powerful antigravity will make wormholes, the renowned gateways in space–time, an everyday reality; though relativity holds that they open and close so fast as to go unnoticed, phantom energy will keep them open. Space travel, the stock-in-trade of science fiction, will then become a reality: spaceships will travel through wormholes as a short cut either to another universe or to some point far back in time in our own one.

So the answer to our initial question on the future of the universe is ambiguous, as if coming from a modern-day oracle of Delphi. At least on the basis of our present knowledge, we cannot rule out either a closed universe that will revert to its former self, or an open one that will constantly expand. And what's more, both cases have several intermediate variations; initial hypotheses often vary, and the slightest change to any parameter in the mathematical models under construction may have substantial consequences.

The fact that scientists are preoccupied with the future of the universe is striking, to say the least. Finding a definitive answer is no easy matter, nor will it be of any practical significance for finite human life. Investigating physical laws to their furthest limits may of course lead to greater understanding of them, and that alone is reason enough. Yet questions on the future of the universe, which if nothing else spans billions of human lives in duration, also have one noteworthy dimension to them: they are the expression of the insatiable curiosity that has been our hallmark since ancient times. They are the epitome of the search for truth, the painting of a masterpiece that may very well never end. In our utilitarian age, that carries its own particular weight. As astrophysicist Nikos Prantzos, research director at the CNRS in Paris, aptly notes: "Our thoughts on the distant future of the universe are of no practical significance. But from a philosophical point of view, it's worth holding onto the idea that the wonderful complexity of the cosmos that surrounds us, and that has been revealed to us by modern science, may merely be a fleeting, insignificant moment in the history of the universe. Yet even if that is the case, it will have been for a "moment" that is so, so precious..."[22]

At any rate, light has one more unique attribute. It is not only the most primordial entity in the universe, but possibly also the most stable. Even the proton, the infinitesimal building block of matter, may decay in the long run. But no such dangers hang over the photon. It stands proud and alone, its deeper nature inaccessible, at the top of the pyramid of energy and matter particles making up the universe. Even if the light of the Sun is lost and the stars go out, photons will forever wander the universe, perhaps as messengers of the past, foes of darkness. At this very moment, as you are reading this book about light, a few hundred primordial photons lie in your hand. They have existed since the creation of the cosmos, and will continue to do so for as long as its life lasts. From the beginning, light has been synonymous with the universe itself.

8

Epilogue: A Farewell to Light

So the time has come to bid farewell. Personally, I feel that the farewell to light is like the one we bid to a great love: He or She will always exist within us, our eyes carry their image. There are, essentially speaking, no farewells. Yet as time passes, and their absence—the absence of light, of love—grows stronger, the meaning and passion of all we have experienced together come to serve as both a melancholy memory and a paradoxical support. Details are lost, colours fade. But what remains are the invaluable moments, which may have been entire centuries rather than moments; and they always speak to us with the same intensity. There is, wise men say, a nostalgia for love. I would join others in saying there is a nostalgia for light, even when the time comes to bid farewell.

At any rate, from my long acquaintance with light, one sense emerges as predominant: while light itself is invisible, it hides the greatest secrets of nature. The quantisation of radiation and the wave nature of particles, the relativity of time and the recession of the galaxies, the unification of interactions and the conversion of matter into energy—and so many other things—are inherent to light and its behaviour. It's just that light guarded these great secrets jealously, and even passionately resisted attempts to reveal them. Besides, via the "uncertainty" inevitably caused by its presence, light itself—that instigator of knowledge!—placed inviolable limits on knowledge.

There is one central idea running through the pages of this book, which I have tried to highlight as much as possible: the idea that light has led science, and thus our understanding of the world, to its greatest conquests; and also that it is our main source of information on the formation and evolution of the stars, as well as on the very creation of the universe. At the same time, light has a leading role in developing and maintaining life, at least as manifested on Earth. Without light neither plant nor animal life could have existed, in which case human existence, the pinnacle of evolution, would never have found a way to emerge and survive.

At any rate, the truth is that at least ostensibly, this book has focused on the scientific dimension of light. So by way of an epilogue, it is worth summing up the cornerstones of its behaviour. While light has preoccupied human beings since ancient times, it came to the forefront of physics after the scientific revolution. Understanding of it has followed an upward trend ever since, albeit with fits and starts. There have been great leaps, but some backtracking too. The interesting thing is that every time an important property of light is revealed, it opens up important and unanticipated prospects for interpreting the world. In fact, we often find ourselves on the threshold of a true revolution.

The general theory of relativity is a typical example. Einstein's axiomatic hypothesis that light maintains a constant speed irrespective of the source or observer's motion led to a revision of the notion of time; and from there to breathtaking consequences as far-reaching as the conversion of mass into energy. Of course, the shining edifice of classical physics was not toppled, but its foundations were reinforced with new methods and materials.

In the case of quantum physics, the new wind has wrought more sweeping changes. Our stable, transparent world has dissolved into clouds of probability and paradoxical quantum behaviours. The world's new foundations are made up of the uncertainty principle, wave-particle duality and electron quantum jumps, while an unexpected wave-form equation stands as its supreme law. Yet as often occurs with social revolutions, the entire quantum revolution had its beginnings in a seemingly innocuous problem: the laws of classical physics were incapable of providing a satisfactory description of the light radiation emitted by a heated body. The bold hypothesis that the radiation consisted of energy quanta fully accounted for experimental results, but at the same time opened up Pandora's box. So it took three decades of ingenious theorizing and experiments to come up with a comprehensive quantum theory. Once again, light had blazed the trail.

Things took a similar course when electromagnetic forces were successfully accounted for on the basis of photon exchange between charged particles. The detailed theory formulated was thus to serve as a model for the other forces. In fact, as the experiment in question triumphantly verified, 'heavy light' exchanged between particles is responsible for weak interactions. Thanks to light, the search for a unified theory encompassing all forces and matter particles made a major stride forward.

So light does not simply blaze the trail. As is the case in life, it also shines the way and guides our uncertain steps. The complex, peculiar microcosm of matter is now comprehensible to us because it let light pass through one of its skylights to illuminate the interior.

Yet if light was a mere guide in revealing the microcosm, the stars, galaxies and nebulae signal their existence almost exclusively thanks to it. An immense variety of electromagnetic waves is picked up by optical telescopes, space devices and special satellites. The macrocosm speaks the language of light. The universe thus unfolding before our eyes is as breathtaking as it is unimaginable, both violent and wondrous.

In fact, it is in one far corner of the electromagnetic spectrum, in microwave frequencies, that the secret of creation itself lies hidden. Discovery of the "cosmic" radiation reaching the Earth from every direction in the heavens confirmed, once and for all, that the universe was created by a superhot Big Bang. Tiny energy "wrinkles" have even been detected in this primordial light, depicting the time when the galaxies were formed. An important moment in the history of the universe was thus dramatically revealed, leading the significance of light to new heights of glory.

As stated at the outset, my main aim as author was to focus on the scientific investigation of light; but that does not represent my whole self. My visible, rational self derives great satisfaction from the progress of science. Besides, I too have made my own contribution to it, however small. But what remains is my other self: parallel and unfulfilled, hidden and torturous. What remains is that voice, at times hushed and always unfathomable, which finds unspoken and yet sharply defined elements in light, aspects of it that are unspoken and yet entirely our own. That self, that voice, tenderly accompanied me as the book was being written, while time raged outside. What might that voice penetrating my past and present have been whispering, what might it have had to add, above and beyond the scientific data?

Even now, as this book approaches its end, I find myself unable to express precisely what the whisper meant. In any case, if it were possible, I believe it would be rather dishonest towards my readers to do so. It might prejudice their judgement and their own quest. A book—every book—has overt sides to it, while others are directed at all the unspoken, often painful things permeating our lives. So the autobiography of light should first be read with the mind, and then either later or in parallel with the heart. Is that not true of all autobiographies, or perhaps all books and their authors? Whatever the case, the more the deeper texture, passions and peculiarities of light are revealed, the more wonder it provokes. Yet within it, that wonder also conceals some unavowed elements of envy.

First and foremost, let's take a good look at the theory of relativity. One of its unanticipated conclusions reflects a human desire that often grows heart-rending: light is ageless, time does not pass for it. This incredible attribute raises light above and beyond everything else in this world, every form of matter and living existence in it. Light alone escapes the silent web of time. All else in the universe—from the microscopic particles of life to the stars in the galaxies—exists and unfolds in time. Time is the supreme ruler and judge of all, both the wondrous creator and the instigator of decay. That fate is unknown to light. Time passes beyond light and its metamorphoses; light neither submits to time nor lives under its threat. Perhaps, then, the attraction towards light I felt as an author had its roots in this stunning uniqueness. Besides, doesn't every great love have that element of uniqueness? Does it not live on in the illusion that it transcends time—that it will exist forever, as it always has done?

Perhaps because it converges with what the soul secretly suspects, one other attribute of light arising from its quantum nature is equally awe-inspiring: two photons that were originally together will always continue to interact. One photon feels the other's influence or change in behaviour instantaneously, even if they are now millions of light years apart. It is as if there were some unknown means of communication between them. Similar things are said to occur when you are truly in love, and sense that your identification with your partner transcends space and time.

In any case, it's worth stressing that despite having been exhaustively investigated, light still resists any precise determination of its nature. What is a photon? Why is the speed of light a universal constant? Why does that constant play a fundamental role in structuring physics? How can we account for the fact that all of nature's interactions mimic the trail blazed by light? What exactly do we mean by saying that light is both a wave and a particle simultaneously?

Nobody knows whether the answers will ever be found. Personally, I suspect that every answer will give rise to profounder questions, and that no final answer exists. Besides, I believe that what our times persistently call for is a synthesis of knowledge rather than the compartmentalisation of it. Knowledge is not the product of specialisation, as our hurried times teach. On the contrary, it is the attempt to discover bridges between the sciences, and to build other solid, lasting bridges between science, philosophy or art. "The ongoing fragmentation of knowledge and resulting chaos in philosophy are not reflections of the real world," notes biologist Edmund O. Wilson, "but artifacts of scholarship. The propositions of the original Enlightenment are increasingly favored by objective evidence, especially from the natural sciences."[1] Unless and until the propositions of the Enlightenment are realised, light will remain the element most unifying the sciences, while also serving as a timeless, shining bridge, in this instance between science and art.

If seen simultaneously and in parallel in all its dimensions, scientific or religious, philosophical or artistic, light acquires a clarity all of its own; even if, as physics insists, some secret still escapes us at its core. Fully deciphering that secret may mean interpreting Everything. That interpretation may someday be propounded. The dogged pursuit of a Unified Field Theory goes on, and a group of equations will one day describe its core. But yet again that will have little impact on the other, torturous side of the secret, which is linked to our very existence itself.

So now that the time has come to bid farewell, I have the suspicion that the doubts surrounding the nature of light are an integral part of its charm. Just as a woman's charm is accentuated by the hidden sides of her self; that momentary glint in her eyes that expresses her innermost feelings.

The more I think of it, it can't have been altogether coincidental that the woman who accompanied me on my constant wanderings—or so I imagined—always smiled enigmatically whenever I tried to guess her intentions or scrutinize her true face. As with light, the attraction I felt was not easy to put into words, yet lurking within it there may have been something of awe or infinite nostalgia. Besides, as time passes, I have the absurd sense that my farewell to light may not be final. After the end, which may also be a beginning, I am once again going to meet some kind of light—though this time total and absolute; and She will always accompany me. Over so many centuries of aimless wanderings, perhaps that's what her enigmatic smile was trying to say.

Heraklion, Crete, October 2005

14 billion years after the Big Bang

This book was nearing completion. All that remained were some additions here and there, a few corrections and then the final proofs. But one night, at the hour I am usually shrouded in fears and dreams, a primordial, universal light made clear its need to communicate. It was intending, as it said, to retreat into silence, and wanted first to say a few things about itself. With some effort I took notes from its confession. And if that sounds paradoxical, or perhaps some sort of delirium on my part, science can confirm the fact: right this moment, dear reader, there are a few hundred primordial photons in every cubic centimetre of your room. In other words, particles of light that witnessed the creation of the universe, and which are now arriving on Earth from every direction in the heavens. They are a fossil of light, yet one free for us to touch and thrill at, for every human being to touch.

Like all fossils, this one yields evidence of a distant past, which in this instance coincides with the beginning of the Cosmos.

The discovery of primordial light was of great importance to cosmology; and we have seen that it happened by chance. In 1965, two physicists at Bell Labs in the USA were using a special antenna to try and pick up signals from a man-made satellite, but their efforts were hampered by some kind of paradoxical radiation reaching the antenna in the form of microwaves. It was then established that the radiation in question was the echo from the mind-boggling Big Bang that created the universe itself over 14 billion years ago. In fact, the presence of that radiation had long been predicted. In science as in life, however, bold ideas often go unnoticed.

It is that primordial light which tells its story here. Only that a professor of physics had the good fortune to record its words and put them down on paper, one blessed night when an unusual silence reigned over the city. He cannot swear that he didn't miss a few words, or even entire phrases. The primordial light had dimmed by then, after a journey lasting so many billions of years. But it had experienced the universe ever since space was created, and ever since time began to run its course. Of course, light itself is ageless; the passing of time means nothing to it. That remarkable attribute never ceased to provoke paradoxical feelings in the author, which may have had something of envy in them. Hurriedly and with great effort, then—and while time raged on outside—he noted down all that light had to relate. He remembered from the old days how, in the Euripidean tragedy Orestes, *Apollo appears as a* deus ex machina: *"I am the light," he announces. "Hear me—hear the light since you cannot see it. The light is an order of truth."*

So it is light, then, that is anxious to have its autobiography—and its truth—heard here.

The author

9

The Autobiography of Light

I WAS BORN CENTURIES UPON CENTURIES AGO, in a Space where there was no space, and at a Time when there was no time. Yet in a strange way, I feel that I pre-existed my birth. And while everything has changed since then, I feel that nothing changes. My presence is the measure of eternity.

In any case, it's not worth discussing things like my birth and the birth of the Cosmos, which have always provoked debate. What bolsters my confidence is the fact that my presence has always seemed to be paramount: in myths or whatever is known as science, in the works of mortals as well as in religions, light has always had a special role to play. From time to time many have wondered about its nature, while others have simply stood in wonder and acceptance. Besides, the indisputable truth is that poets and artists have sung the praises of light more than anything else.

So the time has come for me, the most ancient, primordial light to talk a bit about myself. I don't know the reason why this need arose in me. Old age, with its fondness for reminiscing, has no meaning for me. Nor, as I have already admitted, do I have any complaints when it comes to honours and recognition. It's just that I'd like others to learn of the miracles and strange things that only I was capable of experiencing. Of course, it's often the case that autobiographies are prompted by other, profounder motives: by the vindication of our actions or ideas. If, as my retelling moves along, I discover that such motives lie within me too, I promise not to hush them up. Besides, that's a premise of transparency, which has always been characteristic of my life—or to be more precise, at least from some point on.

Transparency! Without meaning to, I've touched on a turning point in my eternal journey, which is quite rightly thought of as being hugely important.

G. Grammatikakis, *The Autobiography of Light*, https://doi.org/10.1007/978-3-031-56917-3_9

Sure enough, 400,000 years after the universe was born, I stopped interacting with matter. As they say, that was when the age of transparency began. Of course, I hasten to stress that whenever years or time periods are mentioned in my retelling, they don't have much meaning for me—they just make it easier for anyone listening. Anyway, ever since I've been set free of my ties to matter, I've been able to relay images and events from the very distant past. For later arrivals, the age of transparency marked the beginning of their acquaintance with the early universe.

BUT LET'S PUT THINGS into some sort of order. The universe is said to have been created by a superheated Big Bang roughly 14 billion years ago. It's also said that I, primordial light, was then the dominant component of the universe, while matter was a trifling admixture. I don't know who is responsible for the phrase "In the beginning was Light", but let me just say in all humility that that's hardly a wild exaggeration.

Anyway, the fundamental components of matter in the early universe were quarks, electrons and other heavier particles, which rapidly decayed. But there was also an equivalent number of antimatter particles. Antimatter doesn't appear to exist in today's universe; it vanished back then, in the constant collisions between particles and their antiparticles. I witnessed that crazy dance, that death waltz between matter and antimatter. All that survived was an excess of matter particles, and mostly quarks. The reasons for the bias in favour of matter are known to science. But for reasons we'll see later, I believe that matter quite simply "had" to exist. At any rate, very soon any surviving quarks began to bond in threes and form protons or neutrons; and they make up atomic nuclei. Please don't take this as an ego-trip, but in actual fact it's worth stressing that ever since then there have been a billion photons for every proton or neutron! So in the superheated phase over the first hundreds of thousand years, the energy of light was far greater than that contained in matter. The universe was in the reign of light.

In any case, nucleosynthesis lasted no longer than 15 min, stopping at simple nuclei of deuterium (heavy hydrogen) and helium. Later on, if I'm still in the mood I'll explain how the heavier elements like carbon, lead, gold and so many others are produced in cataclysmic stellar explosions.

But I've got carried away chatting, and one vital fact has slipped my mind: one consequence of the Big Bang that created the universe was constant, break-neck expansion. From being as tiny as a cosmic egg, it soon assumed staggering dimensions. And together with expansion, the temperature also began to fall. Every time the universe doubled in size, its temperature halved. So although it had initially been incredibly high, one hundred seconds after the Big Bang the temperature was no more than a billion degrees. This constant cooling directly impacted my own frequency, which shifted from the visible area to radio waves. But what's really important is something else: today, even though 14 billion years have passed since its explosive beginning, the universe is still continuing to expand. Since the energy of the Big Bang is sufficient to overcome gravitational attraction, the galaxies are receding from each other at enormous speed. This recession is betrayed by the light from galaxies itself—that's just how revelatory the role of light always is!—and confirms what I experienced at first hand: that the beginning of everything was an unimaginable Big Bang, which via complex processes in space and time created the cosmos of the seen and unseen.

No doubt those listening to me will already have serious doubts as to how genuine or, let's say, how accurate my memory is. I too have realised that my account seems unbelievable, a little like a fairytale. But the same is true of some stories about life, or at least some of the hidden routes it takes. As far as I'm concerned, I can confirm that my autobiography is based on facts. Yet in every autobiography—to say nothing of a potted one—many things have to be left out, and there are others it's not right for me to touch on. So as to the reasonable question of what existed or what happened before the Big Bang, I'll stick to the answer given by science. It is, so the argument goes, a meaningless question. Space and time were created together with the Big Bang, and are inconceivable as absolute entities; so "before" does not exist either. As for the even profounder question of what caused the Big Bang and what purpose it served, allow me to remain silent. Not that I don't know. It's just that I'm in an awkward position, and some things cannot even be hinted at.

AT THIS POINT I think a word should be said about the forces acting on the early universe. I won't insist on my own personal testimony—which is of course unique—so as not be seen as prejudiced or one-sided. Besides, science's opinion, which goes by the name "Unified Theory", appears highly compelling. As its name indicates, it tries to encompass in one framework all of the forces or fundamental particles that had a definitive role to play in the first moments of creation.

In today's universe there are only four of these forces, which are often also called interactions. The most familiar one is gravity, which dominates the macrocosm and determines the movements of the heavenly bodies. While it appears extremely strong, it is the weakest of all. Though it could hardly be otherwise, the force holding protons and neutrons tightly bound into atomic nuclei is by contrast very powerful. It is to this "strong" force that we owe the stability of matter. The third or weak interaction apparently has the oppo-site intention: it breaks nuclei down, releasing radiation, and is generally dominant in nuclear reactions.

Last of all comes the electromagnetic force, which holds electrons in orbit around the nucleus. In other words, it is what binds atoms together. Also in its sway are electromagnetic waves, which extend over a broad spectrum of frequencies. It is, quite simply, the flip side of light. It obviously sounds strange that light has wave properties in addition to its particulate existence, but that is our deeper nature, and what at some point led to my being discovered. We should note that a dual nature doesn't appear to be all that uncommon; plenty of people reveal an unknown side to themselves when conditions permit.

So the four interactions start out from the microcosm of matter, and spread like an enormous web over the entire expanse of the universe. This web is exceptionally dense at atomic dimensions. There, three of the forces inter-twine via thin, incredibly strong threads. Gravity is left to dominate alone in the macrocosm, so it's hardly strange that it is the driving force behind the formation of galaxies and stars.

All that now remains is to clarify whether these same forces were also present in the early universe. Scientific opinion is categorical on this, and corroborates my own testimony. At the extreme energy and temperature conditions then prevailing, the four interactions were joined together, so a unified form of force was responsible for the first moments of cosmogony. All the same, I have to admit that since there was no clear-cut distinction between time and space, my memories are somewhat hazy too.

This period, characterized by a single unitary force, was infinitesimally short. Gravity soon separated from the other three forces, which remained united. The cosmic fluid was made up of quarks, leptons, neutrinos and their antiparticles. And of my ubiquitous photons, of course. The interactions were frenetic. Soon strong force broke away from the others, too. This was the time when antimatter disappeared. After complex processes, any quarks not sacrificed in catastrophic collisions with antiquarks would go to make up the matter in stars and planets.

All the same, the universe never ceased to expand and cool. At some point, once conditions were right, the final bitter farewell came: weak force sepa-rated from electromagnetism. The four forces were now distinct, and each took on its own special role in the evolution of the universe. Incredible though it may seem, less than a second had elapsed since the Big Bang. But the elements necessary for the cosmic drama, which would go on evolving for centuries upon centuries, were already there: the protons, the neutrons and the electrons, as well as the four basic interactions. Time and space, the stage set for the drama, had been created hand in hand with its beginnings.

SO IN THE EARLY STAGES of the universe, the wild, uncontrolled activity of its primordial elements—to which I was always witness or partner—led to the synthesis of protons, and later to simple deuterium or helium nuclei. At some crucial point, however, the intervention of electromagnetic force trapped the similarly abundant electrons in stable orbits around protons. So 400,000 years after the Big Bang, stable, electrically neutral hydrogen atoms were created. Hydrogen has abounded in the universe ever since, making up the greater part of its visible matter.

But those are not the only reasons why I described the creation of atoms as crucial. It has also proved crucial in my own personal life. All of a sudden, I lost my energy supremacy. This was due to the fact that as the universe expands, the photons I consist of cool down and constantly lose energy. So although there are always a billion times more photons than protons and neutrons, the latter's stable energy has surpassed my own; from being dominated by radiation, the world has been led to domination by matter. It would be lying to say that this development did not irritate me. In fact, if I wanted to be ironic I could say that we have passed from the era of photocracy (the reign of light) to that of materiocracy (the reign of matter). But a comment like that wouldn't exactly show a sense of responsibility. All of us who took part in the evolution of the universe—especially me, as the protagonist—know that every change is subject to more general laws. Whether those laws are imposed by chance or some necessity is an endless debate, which I for one wouldn't want any part of.

All the same, there is a deeper reason why the creation of atoms also determined my own course in life. Since atoms are electrically neutral, I ceased to interact with matter. In a sense I, primordial light, had gained my liberty, and could now travel free of any material commitment. I was to conquer every corner of the universe, experiencing the toughest and most creative of times. I was also to transmit useful information to anyone able or willing to receive it. History, they say, has its written monuments; without exaggerating, the moment atoms were created and I was set free, the universe passed from prehistory into its own history. So my release from matter makes direct observation of the past possible.

THE BEGINNING OF THE HISTORICAL ERA—400,000 years after the Big Bang—was thus marked by the formation of atoms, and the ever more obvious domination of matter.

Having secured an existence of sorts, the universe doesn't seem to be in any hurry. Gravity has gradually come to dominate the macrocosm, while the other three forces have limited their activity to the atomic world. As for me, I am enjoying my freedom; and the fact of the matter is that no-one can match me for speed when it comes to moving and conquering space. In fact, as time has told, no material body or other radiation is permitted to reach that speed, let alone exceed it. The only thing that concerns me is that as the universe expands and cools, my photons are also cooling; in other words, they're losing their energy. But I'm convinced that this too is part of the laws which, yet again by chance or necessity, govern the evolution of the universe.

To sum up, then: following the first phase of creation, the universe was composed of an enormous cloud of hydrogen and small amounts of helium. Under the influence of gravity, minor changes began to occur in the cloud. Some random accumulation of matter in one area acted as a magnet for other atoms. This constantly grew in strength until at some point, as in an avalanche, the accumulation gained appreciable density. So the originally uniform cloud now had scattered lumps of matter. These were the proto-galaxies, resembling islands in an indifferent ocean of sparse matter. As I immediately suspected, they were destined for a major, dramatic development. The truth is that I had to wait a few more billion years for it to happen, but I don't regret it and I'm not complaining. Not only because the universe changed shape from then on, but also because I found company on my endless journey.

Let's put things into some sort of order again. The universe was now dotted with primary galaxies. It had reached half its present size, and its temperature was now only a few degrees above absolute zero. The uniformity of the large primordial cloud had already been violated by the protogalaxies, but then the same thing was to repeat itself in each protogalaxy, where scattered concentrations of matter were created once more. These were the ancestors of the stars. But gravity continued to compress atoms, leading protostars to contract and thus increase in temperature. At some stage the temperature reached the point where hydrogen nuclei began to fuse: in other words, the onset of an uncontrolled nuclear reaction. Compressed matter self-combusted, releasing enormous quantities of energy and light. In brilliantly betraying its existence, a true star was born!

Despite all the years that have passed, the truth is that the mere recounting of these special moments moves me. The birth of a star, which has been repeated billions upon billions of times, always retains great charm. It lends brilliance and some impalpable purpose to the universe. Besides, in the endless, icy night it adds a bright companion, a lighthouse, let's say, to my own journey, which has lasted aeons upon aeons.

AS A TRUTHFUL, OBECTIVE WITNESS, what I'd like to stress is that the stars were not born once and forever. They are constantly being born, evolving and finally disappearing, on a time scale that varies from millions to billions of years. In fact, the smallest stars live much longer! So stellar generations overlap, and new stars are born as long as there is available matter in the primordial cloud. It goes without saying that starlight is of the same nature as my own, and travels at the same colossal speed—we only differ in terms of frequency.

So stars of every shape, size and age are to be seen in the billions of galaxies that have gradually taken up the universe. But there is one dramatic truth: a star's life is a constant struggle against gravity. The force that gave birth to it has a constant tendency to crush it. To cope with gravitational crushing, the star resorts to ever more complex nuclear reactions. That way abundant energy is released, counterbalancing gravitational pressure. In the first phase of its life, nuclear burning of hydrogen forms deuterium and then helium. The star stops shrinking, and its size stabilises. A very large number of stars have reached this "mature" age, which lasts millions of years. Sirius, the Pole Star and Vega all come to mind, for some reason.

Yet as experience goes to show—and I've got plenty of it!—nothing lasts forever. At some point the hydrogen in the star's interior runs out, and the threat of gravitational crushing looms once more. The star is compelled to move on to new nuclear combustion. Helium is now burned to produce the important element carbon. But this time the energy is released at a rapid pace, and massive expansion is the star's only option. So it is transformed into a "red giant", making a fantastic show as it emits reddish light. The enormous temperatures in a red giant's interior enable the creation of ever more complex nuclei, which in turn take part in combustion. Tragic though it sounds, stars even use the ash from previous nuclear burnings in their desperate defence against gravity. By synthesizing complex nuclei—of oxygen, silicon, iron—yet another step in the organization of matter is achieved. Whatever the complex Big Bang mechanism failed to create is born—or perhaps has to be born—in the interior of the stars!

As I can confirm, all stars either have passed through the red giant stage or will do so. As this doesn't last long, I have only ever meet a few red giants on my endless wanderings. Yet again, I have my reasons for mentioning Aldebaran in the constellation of Taurus, and also Antares, a conspicuous red dot in Scorpio. Even the Sun—a star that'll turn out to be central to my account—will not escape the bitter cup. In roughly five billion years it too will be transformed into a red giant. Its dimensions will grow so much that it will smother the nearby planets and torch them, despite the fact that they have been its faithful companions. However distant that prospect is, it troubles me. It's not just that I have been taught to regard ingratitude as base behaviour. For reasons which will emerge later on, torching the planets, and especially one of their number, will have untold consequences for my own posthumous reputation.

Painful as it is, a star's transformation into a red giant is essentially a precursor to death. It seems that some kind of death or end is an inescapable fate in the universe. At some point, not even the universe itself will avoid it! So however eternal they may seem, stars too have their own death, their own humble or spectacular end. Sure enough if it's young, once a star's nuclear fuels run out, it turns into a white dwarf with an extremely bright, hot surface. But since white dwarfs constantly emit energy, which is not made up for by nuclear burning, they grow dimmer and dimmer until their light goes out for good. I often bump into these dark remnants of stars, which are cold and invisible. How could even the most trusting person ever be persuaded that they once had such a brilliant, interesting past? All the same, I fear that's a more general truth. Nonetheless, all of us hope in vain that the painful moment when we are no more than a shadow of our former selves will never come.

A different fate is in store for larger stars that exhaust their nuclear fuels. Gravitational collapse leads matter to an unusual state whereby its protons are transformed into neutrons, but due to compression, the star then becomes small and tremendously dense. It is now a neutron star, revolving at dizzying speed. In fact, I was deeply moved to discover that a neutron star steadily emits a beam of radio waves. So it's like a revolving lighthouse in space, with very regular signals. Only those who, like me, have experienced the boundless silence and vast distances that dominate the universe can appreciate the value of such a presence. It insistently reminds us of passing time and the everlasting need to communicate with those near us.

THE TRUTH IS THAT from time to time, the monotony of my wanderings is interrupted by another presence, though this time a sudden and blinding one. What I mean is the explosive death of large stars. In the final analysis, ending up as a white dwarf or neutron star can be seen as a calm, dignified end. But for really large stars a darker prospect lurks, involving the cancelling out of their very matter: as if wanting to avoid this bitter fate, such stars eject a large part of themselves out into space.

People arriving on the scene later than me gave the name supernovae to these cataclysmic eruptions, which shine like a billion stars. What causes a star to explode is its eventual inability to prevent gravitational crushing, following repeated cycles of nuclear burning. The star's core then begins to collapse and disintegrate. A flood of neutrons is the first to leave the disaster scene, at a speed close to my own. And as the core density continues to rise, the star's outer layers are blown violently out to create a rapidly spreading, brilliant nebula. The stunning nebula in Cygnus, an endless source of wonderment for me, is all that remains of a supernova explosion many centuries ago.

The time has now come for me to keep one of my promises, and explain how the heavier elements often encountered in stars and planets came to be there. Careful listeners will already be suspecting what the answer is: matter from a star that explodes and expands rapidly in space also contains traces of the heavier elements forged by nuclear burning in its interior. In fact, thanks to the extreme conditions created by the cataclysmic explosion, even more complex nuclear reactions occur, so elaborately structured nuclei such as silver, gold or uranium make their appearance. From then on they too are ingredients in the cosmic fluid.

What this means—and here I'd like your closer attention, please—is that new, second-generation stars formed from the cosmic fluid will contain a variety of heavier elements. Not simply helium and hydrogen, as was initially the case. One second generation star is the Sun, which provoked my interest from early on. As far as I can remember it was formed ten billion years after the Big Bang. So it contains a mixture of heavier elements, whose presence will prove valuable at some point.

The fact is that the universe is an enormous stage, where endless cycles of births and stellar deaths occur. For those who insist on looking for a purpose in everything, ejecting enormous amounts of stellar matter is neither wasteful nor superfluous. This matter has undergone constant nuclear processing and contains elements that are impossible to synthesize in any other way. It is the beginning of a process vital to the emergence of perhaps the most paradoxical phenomenon in the universe: the phenomenon of life.

The truth is that the mere mention of the word "life" awakens vivid memories and moves me deeply. So I have to rein myself in, and get back to the explosions of large stars. When they occur, not all of the star's mass is hurled out into Space. Some vestige remains, and that will end its life calmly as a neutron star. But there is one further possibility: when a star is really large, every line of defence in the core surviving from the explosion proves powerless against gravitational crushing. Gravity even overcomes the apparently uncompressed neutron sphere, leading matter to a state of effective non-existence. Both literally and metaphorically, what is formed is a black hole.

I hope it doesn't sound spiteful when I confess my profound dislike of those bodies, which are only stellar by name. In fact, I'm overcome by the entirely justifiable fear that I might bump into them on my wanderings. And that's because while black holes are invisible and non-material, the gravitational field in their vicinity is enormous. Nothing crossing a forbidden zone called the event horizon can possibly escape. Whether it's a material body or radiation, from then on it remains prisoner to the black hole, incapable of communicating with the outside world. I have seen entire stars heading towards black holes and disappearing, as in the constellation of Cygnus, for example. Even I, light itself, synonymous with freedom and the open life, can be imprisoned by their incredible gravitational field. Hence my fear, and my constant attempt to avoid them. Luckily enough, their invisible presence is betrayed by the anomalous movements caused to nearby stars, or by the enormous amounts of energy released as they swallow any matter approaching their event horizon. From the effects I have observed, I even suspect that enormous black holes lurk in the central regions of every galaxy. It may well also be the case that the quasars detectable at the limits of the present universe, which shine as brightly as a thousand galaxies, owe their terrific energy to a gigantic black hole formed at their centre. Forgive me for being prejudiced, but I feel that revulsion and hatred are the only thing black holes are worthy of. Aren't such feelings entirely justifiable whenever someone threatens our existence and life prospects?

GETTING WORKED UP DOESN'T SUIT ME, though, even over black holes. It's time I got back to my story. I have already stressed that over the course of billions of years, stars are constantly being born, developing and dying. So the wide variety they display is hardly strange. But because the distances between them are enormous, the stars seem very lonely! Even I, light, being so proud of my speed, take decades or even centuries to go from one star to another, and even a hundred thousand years is not long enough for me to cross a galaxy. So however little time matters to me, and what people call age holds no meaning either, I sometimes feel tired from my travels. In contrast to stars, galaxies are relatively short distances apart. They are more sociable than the stars! Another impressive thing is their variety of shapes: some galaxies are elliptical, other spiral ones look stunning from afar, and yet others are irregularly shaped.

One thing is for sure: despite its apparent calm, cosmic space is often violent. So there are plenty of things to distract me on my endless wanderings. Sudden explosions, stars of variable brightness, jets of ionised gases and matter processes beyond conception are standard fare. I have also been witness to cataclysmic collisions between stellar bodies. Only a few year ago, for instance, a comet crashed into Jupiter with tremendous momentum. Even two galaxies may collide; their violent distortion gives rise to impressive phenomena. Added to this, various forms of radiation criss-cross the vast expanses of the universe or are emitted by stellar bodies and make amazing displays. Most are electromagnetic in nature, i.e. relatives of mine. I am especially fond of neutrinos. They have either accompanied me since the cosmos was created or have escaped from the explosive interior of stars. Their mass is infinitesimal, and their speed is close on mine, though of course it can never catch up. Besides, neutrinos have one attribute that sometimes makes me jealous: they are so penetrative that in practical terms nothing can stop them. With minimal losses, they are capable of passing though several planets in a row! Last of all, as I've already mentioned, the comets are common space travellers. Though they are no more than dirty snowballs, they have tails that shine spectacularly in starlight. It seems that illusions of their kind are not that uncommon: deep down, a phantasmagorical appearance often hides an ordinary, insignificant self.

Let's just note that in recent years science has come to be dominated by the belief that a large part of the matter in the universe leads a dark existence. It neither emits light, like the stars, nor can it be seen by any other means, and so it's quite rightly called "dark" matter. Its gravitational effect is nonetheless evident in the motion of the stars and the structure of the galaxies. Yet while its presence is regarded as necessary, little can be said of its identity. Is it unknown particles, planets, or some entirely new, unusual form of matter? Needless to say, I know the answer, but I have an intense dislike of concerning myself with entities that appear to scorn light, and which take pride in their gloomy nature and eternal darkness. So I think the best stance, and one I always recommend in such instances, is disdain. Any conciliation may be seen as a retreat from our own shining values. And don't let this stance of mine be seen as bitterness. It's simply a defence of ideas and the need to respect light and its symbolisms.

That being said, my credibility forces me to return to celestial bodies whose existence is certain, and which are permanent companions of the stars. I mean the planets. The ones I meet on my travels are without number. And a few look very special, as they are illuminated by the light of the stars. Planets have no light of their own, since nuclear combustions never took place in their interior, but they are of the same origin as the stars. As a star forms, the heated sphere of gas in its centre is surrounded by disks of swirling matter. Fragments of greater or smaller mass break away, and over the course of time become planets. Gravitational pull forces these celestial bodies of rock or gas to wander in silence around the star. Asteroids are likewise of similar composition: they too are rocky remnants of the star's formation, though much smaller than planets. My path is crossed by thousands of asteroids, though few are worth any attention. But one asteroid which made a real impression on me was called Eros, as I later found out. Back then I still didn't know the meaning of the word, or the passion and feelings it evoked.

TIME FOR ME TO GET TO what's perhaps the essence of my tale. At some point, for reasons I was to understand much later, a new star drew all of my attention. Its composition had been enriched with heavier elements from supernova explosions, while several planets that had also formed from the same primordial nebula began revolving around it. I won't go into detail, but the fact is that after all sorts of processes and re-alignments, an interesting celestial system with a bright star at its centre was created in one of the galaxies. Much later on (of course!), it was named the Solar System.

The Solar System represented an advanced level of stellar organisation. In total, eight planets of varied composition and different size travel in elliptical orbits around a blazing star called the Sun, from which they receive energy and light radiation. The Solar System also includes several satellites revolving around the planets, and is criss-crossed by innumerable small meteorites, as well as comets with showy tails. Yet neither the texture of the Solar System nor the harmony of the planetary orbits could account for the ever-growing interest it provoked in me. I sensed that something special was going on here; that in some strange way the universe was creating an environment so as to enable a difficult, major leap to take place.

It came as no surprise, then, when I realised that one of the inside planets seemed to be tracking a course all of its own. Its being the right distance from the Sun certainly helped, since it allowed mild temperatures to take hold on its surface. This planet, which was much later to be named Earth, had definitely also known a turbulent past, with intense volcanic activity, geological upheavals and electrical discharges in its early atmosphere. But now, five billion years after breaking away from the astral nebula, it had attained relative equilibrium. It had an atmosphere rich in oxygen and nitrogen, oceans and dark-hued continents, with the characteristic green of plants visible from afar. Without any doubt, it was a planet typified by variety and considerable peculiarity. It thus appeared to be burdened by fate, an inescapable fate that would lend an entirely different meaning to the evolution of the universe.

I'm in no position to say when this fate, which is called Life, began to take shape. Nor can I describe all the delicate stages that followed once, by chance or necessity, the chemical molecules of life had formed. But what I do want to stress is that the phenomenon of life resembled a frisson—a persistent, unprecedented shiver that substantially changed the face of the planet via routes both hidden and plain to view.

THE EMERGENCE OF SINGLE-CELLED ORGANISMS is regarded as the first stage of life on Earth. Since then, a wondrous, lengthy process has led to the synthesis of its most advanced forms. As I too can vouch, the presence of life was initially limited to lakes and oceans. Some kinds of jelly-fish and fish may have been its earliest forms. When life spread onto land a few hundred million years ago, the Earth was still in the throes of incessant volcanic and geological activity. Today, now that the planet has stabilized into the image familiar to us, with its seas, rivers and mountains, an incredible variety of living organisms live everywhere, and verdant forests, flowers and grass cover a large part of its surface. I see no reason to hide the fact that I often catch myself marvelling at the beauty of the Earth, which the blossoming of life has lent a different meaning to.

To pre-empt queries and objections, I'll now make a necessary digression. My listeners—and like any storyteller I hope I have a wide circle of them—may well be asking themselves and others an important question: are there no other worlds characterized by complex or at least primitive forms of life? I have no difficulty in confessing that I, and perhaps I alone, know the answer. Since the birth of the Cosmos I have been wandering among the stars, and the planets I have encountered are without number. So I am in a position to know whether life now exists anywhere else, whether any of the planets are inhabited, and how well developed or different that other life is. But there is no way I'm going to make that knowledge of mine public. Not because, as with other things, it is forbidden. It's just that I'm afraid my answers wouldn't be believed! The truth is often harsh and unanticipated. Yet harsher still is the position of those who know the truth and have chosen silence.

But let me return to life, which on planet Earth found warmth and hospitality. It's notable that over the course of time many life forms have disappeared, while those to survive have shown ever greater resilience. This seems to be a fundamental rule of evolution. At times, however, the extinction of a species is also due to outside reasons: one such dramatic extinction was caused by a meteorite crash roughly 65 million years ago. I witnessed that terrifying event from the sidelines, and obviously hope it is never repeated. Yet there was—or maybe there had to be—a beneficial consequence of it. The dinosaurs, supersized reptiles that had dominated the land and air up to then, were almost wiped out. Certain species of birds or reptiles also had the same fate, as did part of the vegetation. But that was when the rapid growth of mammals began. They adapted easily to the environment, and so many ever more advanced species constantly appeared. Among their number—to mention only the best known—we can make out the ancestors of the modern horse, elephant and whale. From that point on I had a hunch that evolution was advancing towards some sort of crowning act.

THAT CROWNING ACT was the emergence of humans. I hope my judgement won't be seen as biased due to my special ties to them. What can't be disputed is that that gifted product of evolution rapidly succeeded in dominating the planet. Humans have colonised almost every corner of it, conquered the sea and air, and subjugated nature exclusively for their own ends.

Anyway, it's difficult to believe that humans share roots with other mammals, and indeed with some species of apes. They don't have the same abilities, nor do they display similar intelligence. Yet common descent does not appear to be disputed any more. It's just that it took millions of years for the human brain and nervous system to evolve to the present level of complexity and organisation. Since then, however, a very profound change has come about in the evolutionary course of humans. Biological evolution has been outflanked by conquests of the mind and experience, which are handed down from generation to generation via language or memory. I must confess that if I hadn't experienced it, I would be hard pressed to believe it. But the fact is that thanks to accelerating cultural evolution, it has taken less than twenty thousand years for tribes of hunters and nomads to become the people of today: artists and scientists, inventors of technological advances with an awareness of themselves and the world.

It's not worth going into detail. The fact is that today, as I recount my own trials and those of the universe, the intellectual supremacy of humans is beyond dispute, and their achievements really are without end. All the same, it's true that this incredible progress has not always been matched by respect towards the planet's environment. Nor do humans care for nature's wondrous creations and the variety of life. For me, this unthinking attitude is a constant source of both sadness and concern. Many times on my endless wanderings I have experienced the Beginning, but often also the End. No celestial body—and this obviously applies to the Earth—has an eternal fate decreed for it, nor do humans have some invisible, everlasting guardian. So here's to hoping that responsibility does not disappear for good, and that the powers of Good and wisdom will prevail on the Earth.

One thing worth your attention, and something directly connected to me, is that from very early on humans have had an innate curiosity to learn about the world and themselves, to understand the Entirety surrounding them. They have created myths and gods to account for physical phenomena and their own fate. And though I risk being called bigheaded, they have always ascribed an important role to light and its thousand faces. They did of course always have in mind the light of the Sun, and that coming from the distant stars. My own presence—and uniqueness—took many centuries to perceive.

What greatly contributed to this was the orderly process of knowledge and testing that was to be called science. Not to boast again, but it's perhaps no accident that science had its roots in an ancient land where light was wondrously clear. A major civilization blossomed there, and for the first time, questions were asked about the cosmos and its interpretation. Still, the growth of science that was to lead to major discoveries and their applications has mainly made its mark on recent centuries. The power of human ingenuity was to grasp many of the things I have seen and experienced on my wanderings, and to investigate the world by persistent observations and experiments. That was how it discovered both the basic laws governing phenomena, as well as their profounder interrelatedness. Humans now know the secrets of matter and of the chemical molecules that make up life. They have studied the stars and planets, even using devices that accompany me in Space.

But human science was to be particularly preoccupied with light. That greatly flatters me, considering that every form of light is a first-degree relative of mine. How then can I, one aspect of light, not feel satisfied that science has been led to its greatest revolutions and milestones thanks to light? Inordinate modesty is sometimes a sign of weakness. So I, the most primordial form of light, have no hesitation in underlining the contribution we have made to deciphering the Cosmos.

In parallel with science, humans also developed a set of knowledge and practices that was to go by the name of technology, which rests on science. Conversely, scientific progress would be impossible without the new techniques and materials being invented. This wondrous interplay has reached its peak in recent decades. That's how the triumphant discovery of me became possible. Besides, I wouldn't have found the courage to tell my story today if I hadn't made a name and reputation for myself back then. Plenty of unknown people speak in wisdom and knowledge, but their words are drowned out by the babbling of the famous.

But I should rein in my own chatter once more, and get back to the universe and its secrets. It seems truly incredible that a being which inhabited a minuscule dot of a planet felt the need to investigate the inconceivable Totality that encompasses my own existence, too. All the same, it is a fact. Indeed, humans have even invented an important instrument, the telescope, which captures the light from the distant stars and galaxies, opening up new horizons in observation capabilities. Complex telescopes have now been installed at many points on the Earth, catching all sorts of electromagnetic radiation from the infinity of space. Still other observation instruments accompany the space modules flying at great altitudes above the Earth. That way, an unimaginable wealth of images and information on the stars and their evolution has been arduously amassed. So thanks to their ingenuity, humans have been led to conclusions and interpretations not far from the truth, or at least the way I have experienced it. They have even set foot on the moon, Earth's inhospitable satellite. It was a difficult, dangerous journey, which enabled them to gain first hand insight into a celestial body while also advertising their technological grandeur.

But the important thing is that humans also concerned themselves with the very creation of the universe, and that at some point they suspected it had its beginnings in a Big Bang, which is still pushing the galaxies into constant retreat. Indeed, some scientists predicted that if the Big Bang theory were correct, then the reverberation of that terrific event must exist: a diffuse luminous radiation, which should be reaching the Earth from all directions. In short, they suspected I had been born, and predicted my presence today!

IT STANDS TO REASON that I should be overcome with justifiable pride and awe at these developments. In fact, suspicion of my existence was so daring that for years it went unnoticed. Besides, not even the idea of a Big Bang was accepted by everyone. An eternal, unchanging universe seemed more reassuring. Until, at some point, the big day came when the human race discovered me, and quite out of the blue no less.

The truth is that the mere memory of that event sends a shiver down my spine. I can remember those moments in every detail. At some point on the Earth, two scientists—blessed be their names—had set up a weird antenna to receive signals from a man-made satellite. But reception was hampered by some stable, inexplicable radiation resembling light noise. Even a few pigeons were sacrificed—and here I must express my grief—because they had nested in the antenna. But all in vain!

However, determining the actual cause of the noise did not take long. It was due to my own eternal, ubiquitous existence. In fact, as I've already mentioned, from early on other scientists (blessed be their names, too!) had predicted my features: I would have to move in radio frequencies and reach Earth from all four corners of the sky. And, lo and behold, the radiation that just happened to be picked up by the antenna tallied perfectly with the predictions.

Of course, that initial confirmation did not prevent even more accurate measurements from being re-taken in various different ways. That's how science works: by constant doubts and checks. But soon everyone was persuaded of the earth-shattering reality (even though it concerns me, how else should I put it?) that I had been discovered.

However inelegant it is to boast, these developments had an enormous impact both on the world of science and on ordinary people. They even led the greatest sceptics to concede that the universe has not always been the same, and that it was created at some point by a superheated Big Bang. In fact, using their knowledge on the structure of matter, where incredible progress had been made, scientists even managed to draw close to the conditions immediately following the Big Bang. And that's how they shaped a history of the universe which, for all its gaps and flaws, was logically consistent and fitted in with all the observations. People even attempted, and are still attempting, to investigate the "cause" of the Big Bang, and often talk of its "meaning". Being the only truthful witness to the Big Beginning, I am not permitted either to comment or to reveal anything. I am just overcome with melancholy because I know the truth, and I appreciate the fact that humans are in such an agonizing quest for it. Yet it is a truth that will forever continue to escape us, however much some people may claim the contrary.

Nevertheless, it was my own undisputed presence that led to a strenuous attempt to investigate my prehistory and nature as far as possible. That's common practice in human affairs, and not always innocent. Some decades ago, humans launched a special space satellite in my honour, named the Cosmic Background Explorer. With its amazing technological apparatus, the satellite made precise measurements of my attributes and confirmed how well they coincided with predictions. In addition, it confirmed that the formation of the galaxies, a crucial moment in cosmic evolution, had left its imprint in my texture. Like wavelets on a vast ocean, the faint "wrinkles" detected by the satellite in my uniform energy field corresponded to timid, early concentrations of matter. I had only just been liberated from constant collisions with matter particles, and scientists' calculations showed that that was when the formation of the galaxies began. In other words, human genius and skill had managed to "see" the universe's early past and palpate one of its most crucial moments. "It's like seeing the face of God," enthused one of the researchers. Being in a position to know more, I smile at such comments, which do no more than display shallowness and arrogance.

THE TIME HAS COME for me to highlight a personal truth. The discovery of my existence was not merely one of the most significant in science. For my own solitary and endless life, it was also of immeasurable importance. It was like being reborn; that other birth of mine, so many billions of years ago, had no reference to space or time. But from the day those blessed people discovered me, I felt that I existed and was of some significance. Not just for the universe itself, which I am after all an inseparable part of, but for all those who have been moved and led to harbour expectations on account of my presence. If truth be told, however much we ourselves know our own value or hidden sensitivities, we are always in need of acceptance from others, we want them to reach out to us, too. Anyone who has lived alone for a lengthy time or lacked a sense of identity in their life will be able to understand my feelings.

At any rate, it's worth stressing that the existence of humans not only led to my own self-knowledge, and not only gave me a name and characteristics. The universe itself gained knowledge of itself that way. As I have mentioned, it began to be described via myths and theories, provoking questions and a sense of awe. However true it is that the Sun appeared a few billion years ago, in some sense it began to exist when the primordial micro-animals felt the warmth of its rays; it acquired an identity when humans, as rational creatures, sang the praises of its beneficial light and began wondering about its nature. I have the suspicion that the existence of everything presupposes some record of it being made. But recording presupposes perception and senses, nervous systems and a mind. Without them, without language that names and invents, the universe would have remained a cold, indifferent lump of matter and radiation. It would resemble some exquisite music never heard by any ear, a deathly landscape where no sob was ever heard, a play whose plot had never been seen.

I have the feeling that I've let sentimentality get the better of me – sentimentality that my position and prehistory should not allow. But right from the outset I made it plain that my account would be frank; and I'm not afraid of making confessions. What I do know is that if humans and their innate curiosity did not exist, my autobiography would have no meaning. It might never have been written.

ALL THE SAME, I FEAR THE TIME HAS COME for me to retreat back into my silence again; to submerge into the darkness of the universe, interrupted only by the light of the stars or the sudden flash of a stellar explosion. Just as I feel that my autobiography has had no beginning in time, nor does it seem to me to have any end. My fate is bound up with that of the universe; my future depends on its future. If, as they say, it continues to expand, I will constantly lose energy and my photons will cool. But if gravity wins out, then at some point the universe will stop expanding. The galaxies will once again begin to draw closer to each other, and my own temperature will begin to increase, along with that of the universe. Since, in the end, all the matter and energy in the universe will condense into a minute, superheated space, a new Big Bang—a new Big Beginning—may repeat the everlasting cycle.

The truth is that these possibilities are of as little interest to me as others that have been put forward. As I round off my autobiography, I sense that I have seen and experienced a great deal; and I am overcome with calm anticipation. It is an uncommon blessing for someone to feel their life has been an endless bright line; whether or how that line will end appears to be of little concern to them. Besides, on my own journey, one invaluable gift counts: the fact that in reality I am ageless, that time does not pass for me. In fact, I am fully aware that this gift may very well be a source of curiosity and envy. Humans, in particular, appear to be haunted by the passing of time; and they resort to all kinds of tricks to escape its shadow. I don't blame them for that, and have every respect for a threat that is unknown to me. To ageless me, events unfold all together; images of the present accompany those of the past, motion exists within immobility, youth passes in tandem with old age. Though my judgement may be considered paradoxical, I'd like to confirm that this way of life has another charm to it. My autobiography is a series of parallel lines, but ones which intersect at infinite points, perhaps at every single point along them. I have the feeling that some gifted humans also seek solace and inspiration in similar infinite parallel lines, but that their intersection points are defined by the fragments of memory and the boundless courses of the imagination.

SO THE TIME HAS COME for me, primordial light, to withdraw into my own silence. I have had a fortune better than anyone else can boast of: I have experienced the birth throes of the universe, and been an integral part of both its Explosive Beginning and its subsequent evolution. And while I am not permitted to reveal either the meaning or process of that beginning, I can say that as century upon century has passed by, there have been more than a few moments when I've felt awed and amazed. Like the time when a new star shone out from a distant nebula; or when a comet with a dazzling tail approached the Sun's back yard. Nor is it easy to forget the blinding light of the quasars, the explosion of some supernova or the neutrinos that pass through planets and stars. I also consider myself very lucky to have experienced the tortuous struggle with gravity that gives birth to a red giant or a neutron star. But one thing is worth rounding off my biography with: nothing can compare to the shiver of emotion I felt when life appeared on that unique planet.

I was not proved wrong. In fact, I have admitted that it was like being born again, since it was through humans on Earth that my existence and its primordial origins were identified. I gained a name and a reputation. In a sense, I became bound up with the fate of human beings and their manifest need to solve the enigmas of the cosmos. So, as I withdraw into my silence, I am duty bound to confess that it is the images of the Earth and its people that I wish to keep me company in my loneliness from now on. I will always have the need to gaze on the Earth's rivers and seas, its precipitous mountains and verdant plains. From early on I learnt to delight in the sun's games with the clouds, and at other times to listen to the whistling of the wind and the pattering of rain.

Yet above all else, it is the wondrous manifestations of life that dominate my thought and senses. I cannot easily forget the way the first organisms developed on Earth, and then the moments when countless species of animals and plants filled its lakes, the waters of its seas and the land. I am always moved when remembering the days when humans learnt how to use fire, and then to till the land and express themselves in writing. And though my age may be the measure of eternity, how many hours have I not abandoned myself to marvelling at art and poetry, and to listening to the music that echoed human yearnings? And then again, I have often felt a sense of satisfaction when the human mind has uncovered the secrets of the cosmos, and recently got as a far as detecting my own existence.

So now, as I retreat into my own silence—forgive me the involuntary sob—it's those images that I want to accompany me always. They will be both my companion and my solace in my loneliness. In fact, this is the right moment for me to cite a scientific truth, which claims that neither energy nor information is ever lost in the universe. That's how I too have learnt of careless talk claiming that for years now, an author has been working on light, and that he wishes to include my own autobiography in his book. In my official capacity I can again declare that I have no objection, providing he respects my words and the keenness for sincerity I have shown.

Besides, now that I am retreating into silence, sincerity demands that I confess a truth concealed by my inescapable solitude: from now on, in my memory or my wanderings, the presence of that woman will always be intense, along with her melancholy gaze and her voice that ripens into song. I never learnt whether she had a name, and few things are known of her life. However—since no information is ever lost in the universe—she is said to have accompanied the author from island to island and from ocean to ocean, over years and centuries. Still others speak of her never-ending transformations, of her many faces, which—like light—nevertheless remained the same. The important thing is that she seemed to have come from a place without borders, and a time without end. Perhaps that's why the way she carried herself was so ethereal and yet earthly, solitary and then again an integral part of a primordial Rhythm. I have already mentioned that the passing of time has no meaning for me; her presence is thus both present and past together, both eternal and fleeting. "The autobiography of light," she whispered when we met, "is my autobiography, too. So for as long as I exist, it will never cease to be written."

Like every other time, she began to retreat once more. Never for a moment will I forget the way her hands danced in the light, while her wonderful eyes seemed to bear the reflection of her world and of the Cosmos. I, light, shrouded her in my breath, and saw a tear roll down her cheek. I now knew well that like me, she came from a Time where there was no time and from a Space where there was no space. And though her presence seemed to be caressing the present, it also encompassed the entire suspicion of the eternal.

Notes

Chapter 1—Light Gambols Through History

1. Blaise Pascal, *Minor Works,* trans. O. W. Wright. The Harvard Classics. New York: P.F. Collier & Son, 1909–14, Vol. XLVIII, Part 2, p. 449.
2. Antiphanes, *The Theogony* in: Edmonds, *The Fragments of Attic Comedy,* Leiden: E. J. Brill 1959, p. 209.
3. Roberto Calasso, *The Marriage of Cadmus and Harmony,* trans. T. Parks. New York: Alfred Knopf 1993, p. 337.
4. Charles Gillispie, *The Edge of Objectivity: An Essay in the History of Scientific Ideas,* Princeton: Princeton University Press 1960, p. 9.
5. Constantine J. Vamvacas, *The Founders of Western Thought – the Presocratics,* trans. Prof. Robert Crist, Springer 2009, p. 22 (his stress).
6. As translated by George Oppen in his *Selected Letters,* ed. Rachel Blau DuPlessis, Durham: Duke University Press 1990, p. 396.
7. C.C.W. Taylor, *The Atomists, Leucippus and Democritus: Fragments: A Text and Translation with a Commentary,* Toronto: University of Toronto Press 1999, p. 9.
8. *Letters and Sayings of Epicurus,* with a translation and introduction by Odysseus Makridis, New York: Barnes & Noble 2005, p. 7.
9. As above, p. 57.
10. *Select Minor Poems, Translated from the German of Goethe and Schiller by John S. Dwight,* Boston: Hillard, Gray and Company 1839, p. 183.
11. Plato, *Timaeus,* trans. Benjamin Jowett. Available at http://classics.mit.edu/Plato/timaeus.html [Accessed on 5 June 2023].
12. As above.

© The Editor(s) (if applicable) and The Author(s), under exclusive license to Springer Nature Switzerland AG 2024
G. Grammatikakis, *The Autobiography of Light,*
https://doi.org/10.1007/978-3-031-56917-3

13. Saint Augustine, *On Genesis* (I.31), trans. John E. Rotelle, New York: New City Press 2002, p. 182.
14. Aristotle, *De Sensu* 437b 15, translated by G.R.T. Ross, Cambridge: Cambridge University Press 1906, p. 49.
15. Leonard Schlain, *Art & Physics: Parallel Visions in Space, Time and Light*, New York: William Morrow 1991, p. 26.
16. W. K. C. Guthrie, *The Greek Philosophers from Thales to Aristotle*, Abingdon: Routledge 2006 [1950], p. 57.
17. Vasilis Kalfas, *Η μελέτη του ουρανού. Φιλοσοφία και επιστήμη στην αρχαία Ελλάδα* [*Studying the Heavens. Philosophy and Science in Ancient Greece*], Athens: Kallipos 2022, p. 110.
18. In David C. Lindberg, *Theories of Vision from Al-Kindi to Kepler*, Chicago: University of Chicago Press 1976, p. 19.
19. In D. Park, *The Fire Within the Eye: A Historical Essay on the Nature and Meaning of Light*, Princeton: Princeton University Press, p. 78.
20. As above, p. 84.
21. Edward Grant, *Physical Science in the Middle Ages*, Cambridge: Cambridge University Press 1977, p. ix.
22. J. J. McEvoy, *The Philosophy of Robert Grosseteste*, Oxford: The Clarendon Press 1982, p. 33.
23. Margaret Wertheim, *Pythagoras' Trousers: God, Physics and the Gender Wars*, New York: W.W. Norton 1997, pp. 49–50.
24. In Arthur Zajonc, *Catching the Light: The Entwined History of Light and Mind*, New York: Bantam Books 1993, p. 55.
25. As above, p. 56.
26. In D. Park, *The Fire Within the Eye: A Historical Essay on the Nature and Meaning of Light*, Princeton: Princeton University Press, p. 168.
27. As above, p. 171.
28. As above, p. 171.
29. In Arthur Zajonc, *Catching the Light: The Entwined History of Light and Mind*, New York: Bantam Books 1993, p. 73.
30. Galileo, 'The Sidereal Messenger', § 1.6., in: M. A. Finocchiaro (ed.), *The Essential Galileo*, Indianapolis: Hackett Publishing 2008, pp. 67–68.
31. As rendered in Giorgio di Santillana, *The Crime of Galileo*, Chicago: The University of Chicago Press 1955, p. 312.
32. Stillman Drake, *Galileo: A Very Short Introduction*, Oxford: Oxford University Press 2001.
33. Stephen W. Hawking, *The Theory of Everything: The Origin and Fate of the universe*, Beverly Hills: Phoenix Books 2005, p. 79.

34. A. I. Sabra, *Theories of Light from Descartes to Newton*, Cambridge: Cambridge University Press 1981, p. 48.
35. Fernand Braudel, *A History of Civilisations*, London: Penguin 1995, pp. 365–366.
36. Wendell C. Beane and William G. Doty (eds.), *Myths, Rites, Symbols: A Mircea Eliade Reader*, New York: Harper & Row 1976, Vol. 2, p. 339.

Chapter 2—Light's Virtues and Vices

1. Edmund O. Wilson, *Consilience: The Unity of Knowledge*, New York: Vintage Books 1998, p. 73.
2. From Paean 9, in Richard Stoneman, *Pindar*, London: I. B. Tauris & Co. 2013, p. 47.
3. "A letter of Mr. Isaac Newton…", *Phil. Trans. February 19, 1672, vol. 6, issue 80, 3075–6, 3083. Available at* https://royalsocietypublishing.org/doi/epdf/10.1098/rstl.1671.0072 [Accessed on 10 June 2023].
4. Italo Calvino, "Without Colours", *The Complete Cosmicomics,* trans. Martin McLaughlin, Tim Parks and William Weaver, New York: Houghton Mifflin 2002, p. 57.
5. In Barry E. Zimmerman and David J. Zimmerman, *Why Nothing Can Travel Faster than Light*, New York: Contemporary Books 1993, p. 41.
6. *The Odyssey* XIX 172–3, trans. Samuel Butler 1900, from The Perseus Project, www.perseus.tufts.edu [Accessed on 5 June 2023].
7. Odysseas Elytis, "Η συναυλία των υακίνθων" ["The concert of hyacinths"] XX, in *Προσανατολισμοί* [*Orientations*], Athens: Ikaros 2007 (1936).
8. In Henri Bortoft, *The Wholeness of Nature: Goethe's Way toward a Science of Conscious Participation in Nature,* Lindisfarne Books 1996, p. 216.
9. Ludwig Wittgenstein, *Remarks on Colour / Bemerkungen über die Farben,* ed. G. E. M. Anscombe, trans. Linda L. McAlister and Margarete Schättle, Berkeley: University of California Press 2007, obs. 102 & 103, p. 29.
10. In Geoffrey N. Cantor, "Weighing light: the role of the metaphor in eighteenth-century optical discourse" in: *The Figural and the Literal: Problems of Language in the History of Science and Philosophy, 1630–1800,* ed. Andrew E. Benjamin et al., Manchester: Manchester University Press 1987, p. 132.
11. Florian Cajori, *Newton's* Principia*: Motte's Translation Revised*, Berkeley: University of California Press 1946, pp. xvii–xviii.

12. As above, p. 638.
13. Paul Valéry, *Oeuvres* vol. I, Paris: La Pléiade 1957, p. 384.
14. J. Bronowski, *The Ascent of Man*, London: BBC Books 1973, p. 233.
15. Leonard Schlain, *Art & Physics: Parallel Visions in Space, Time and Light*, New York: William Morrow 1991, pp. 69–70.
16. Christiaan Huygens, *Treatise on Light*, trans. Silvanus P. Thomson, London: Macmillan and Co. 1912, p. 4.
17. Charles Gillispie, *The Edge of Objectivity: An Essay in the History of Scientific Ideas,* Princeton: Princeton University Press 1960, p. 420.
18. Henrich Hertz, "On the relations between light and electricity" in: *Miscellaneous Papers*, trans. D.E. Jones and G.A. Schott, London 1896, p. 326.
19. Hermann Bondi, *Relativity and Common Sense: A New Approach to Einstein*, London: Heinemann 1965, p. 60.
20. In Walter Jerrold, *Michael Faraday: Man of Science*, London 1893, p. 20. Available at https://onlinebooks.library.upen.edu/webbin/book/lookupid?key=ha011679000 [Accessed on 12 May 2023].
21. "A dynamical theory of the electromagnetic field" in: Sir William Niven (ed.), *The Scientific Papers of James Clerk Maxwell*, Cambridge: Cambridge University Press 1890, Vol. 1, p. 527.
22. E. N. Economou, $H\,\varphi\upsilon\sigma\iota\kappa\dot{\eta}\,\sigma\dot{\eta}\mu\varepsilon\rho\alpha$ [*Physics Today*], Heraklion: Crete University Press 2003, vol. 1, p. 65.
23. Richard Feynman, "Electromagnetism" in: Fenyman R., Leighton R. and Sands M., *The Feynman Lectures on Physics, Volume II: Mainly Electromagnetism and Matter*, Caltech HTML Edition at https://www.feynmanlectures.caltech.edu/II_01.html [Accessed on 14 June 2023].
24. "A dynamical theory of the electromagnetic field", as above, p. 580.
25. "Action at a distance", in: Sir William Niven (ed.) *The Scientific Papers of James Clerk Maxwell*. Cambridge: Cambridge University Press 1890, Vol. 2, p. 322.
26. Albert Einstein, "Maxwell's Influence on the Development of the Conception of Physical Reality", in: J. J. Thomson, Max Planck et al., *James Clerk Maxwell: A Commemoration Volume 1831–1931*, Cambridge: Cambridge University Press 1931, pp. 66–67.
27. A. A. Michelson, "Comparison of the Efficiency of the Microscope, Telescope, and Interferometer", in: *Light Waves and their Uses* (Lectures at the Lowell Institute, 1899), Chicago: University of Chicago Press 1903, p. 24.

28. Quoted in numerous sources. Originally in A. A. Michelson, "Some of the Objects and Methods of Physical Science", *University of Chicago Quarterly Calendar* 3, 1894, p. 15.

Chapter 3—Light's Revelation of Time and Space

1. Albert Einstein, *Ideas and Opinions*, New York: Random House 1954, p. 226.
2. Alan Lightman, *Einstein's Dreams*, New York: Warner Books 1994, p. 4.
3. Albert Einstein, "Notes for an autobiography", trans. P. A. Schlipp, *The Saturday Review of Literature*, November 26, 1949, p. 11.
4. Arthur I. Miller, *Einstein, Picasso: Space, Time and the Beauty that Causes Havoc*, New York: Basic Books 2001, p. 193.
5. In Peter Conrad, *Modern Times, Modern Places*, New York: Alfred A. Knopf 1999, p. 577.
6. Joseph Schwartz & Michael McGuinness, *Introducing Einstein*, Thriplow: Icon Books 2005, p. 92.
7. Leonard Shlain, *Art & Physics: Parallel Visions in Space, Time and Light*, New York: William Morrow 1991, p. 132.
8. Alan Lightman, *Einstein's Dreams*, as above, p. 33.
9. Florian Cajori, *Newton's* Principia: *Motte's Translation Revised*, Berkeley: University of California Press 1946, p. 6.
10. In Leonard Shlain, *Art & Physics*, as above, p. 132.
11. In Jamie Sayen, *Einstein in America: The Scientist's Conscience in the Age of Hitler and Hiroshima*, New York: Crown Publishing 1985, p. 130.
12. "Relativity (Limerick)", *Punch*, 19 December 1923, p. 591.
13. Alan J. Friedman and Carol C. Conley, *Einstein as Myth and Muse*, Cambridge: Cambridge University Press 1989, p. 57.
14. Lewis Carroll, *Alice's Adventures in Wonderland*, Chapter 7. Available at http://www.cs.cmu.edu/~rgs/alice-VII.html [Accessed on 12 June 2023].
15. Hermann Bondi, *Relativity and Common Sense: A New Approach to Einstein*, London: Heinemann 1965, p. 108.
16. "So Many Constellations", in *Paul Celan: Selected Poems*, trans. Michael Hamburger, Harmondsworth: Penguin Books 1972, p. 68.
17. Richard Gott, *Time Travel in Einstein's Universe: The Physical Possibilities of Travel Through Time*, New York: Mariner Books 2002, p. ix.
18. Jorge Luis Borges, *Ficciones*, trans. Anthony Kerrigan, New York: Grove Press 1962, p. 100.

19. Albert Einstein, "Does the inertia of a body depend upon its energy content?" translated from the original in *Annalen der Physik* 18 (1905) 641. Available online at https://einsteinpapers.press.princeton.edu/vol2-trans/186 [Accessed on 16 June 2023].
20. Roland Barthes, *Mythologies*, trans. Annette Lavers, New York: The Noonday Press 1972, p. 69.
21. Stephen Hawking, *The universe in a Nutshell*, London: Transworld Publishers 2001, p. 13.
22. Pantelis F. Ikonomou, *Global Nuclear Developments: Insights from a Former IAEA Nuclear Inspector*, Berlin: Springer 2020, p. 165 (his stress).
23. From the 1922 Kyoto address "How I created the Theory of Relativity". Quoted in numerous sources, including John S. Rigden, *Einstein 1905: The Standard of Greatness,* Cambridge and London: Harvard University Press 2005, p. 5.
24. Fyodor Dostoevsky, *The Brothers Karamazov*, trans. Richard Pevear and Larissa Volokhonsky, London: Vintage Classics 1992, p. 235.
25. In Alice Calaprice and Trevor Lipscombe, *Einstein: A Biography*, Westport: Greenwood Press 2005, p. 68.
26. John Archibald Wheeler with Kenneth Ford, *Geons, Black Holes and Quantum Foam: A Life in Physics*, New York: W.W. Norton & Co. 2000, p. 235.
27. Subrahmanyan Chandrasekhar, *Eddington: The Most Distinguished Astrophysicist of His Time*, Cambridge: Cambridge University Press 1983, p. 30.
28. Alfred North Whitehead, *Science and the Modern World*, Cambridge: Cambridge University Press 2011 [1926], p. 13.
29. Widely quoted, originally in Ilse Rosenthal-Schneider, *Reality and Scientific Truth: Discussions with Einstein, von Laue, and Planck*, Detroit: Wayne State University Press 1980, p. 74.
30. Marcia Bartusiak, *Einstein's Unfinished Symphony*, New Haven: Yale University Press 2017, p. 131.
31. National Science Foundation press conference "First Gravitational Waves Detected", 11 February 2016, viewable at https://www.youtube.com/watch?v=aEPIwEJmZyE [Accessed on 18 June 2023].

Chapter 4—The Quantum Realm of the Microcosm

1. Victor Weisskopf, *Physics in the Twentieth Century: Selected Essays*, Cambridge: The MIT Press 1972, p. 73.
2. Bertrand Russell, *An Outline of Philosophy*, London: George Allen & Unwin 1970, p. 107.
3. As above, p. 107.
4. Albert Einstein, *Einstein's Miraculous Year: Five Papers that Changed the Face of Physics*, ed. John J. Stachel, Princeton: Princeton University Press 2005, p. 178.
5. A. Einstein, "On the development of our views concerning the nature and constitution of radiation" (1909) in: *The Collected Papers of Albert Einstein* (vol. 2), trans. Anna Beck, Princeton: Princeton University Press 1989, p. 379.
6. From a letter to M. Besso (1954), in Ralph Baierlein, *Newton to Einstein: The Trail of Light*. Cambridge: Cambridge University Press 1992, p. 173.
7. Michael I. Sobel, *Light*, Chicago: The University of Chicago Press 1987, p. 207.
8. Richard Feynman, "Atoms in Motion" in: Feynman R., Leighton R. and Sands M., *The Feynman Lectures on Physics, Volume I: Mainly Mechanics, Radiation and Heat*, Caltech HTML edition at https://www.feynmanlectures.caltech.edu/I_01.html [Accessed on 14 June 2023].
9. In Bernard Pullman, *The Atom in the History of Human Thought*, Oxford: Oxford University Press 1998, pp. 262–263.
10. In Walter Isaacson, *Einstein: His Life and universe*, New York: Simon & Schuster 2007, p. 325.
11. In Ian Duck and E. C. G. Sudarshan, *100 Years of Planck's Quantum*, London: World Scientific Publishing 2000, p. 249.
12. In J. Bronowski, *The Ascent of Man*, BBC Books 1973, p. 340.
13. Victor F. Weisskopf, "Niels Bohr a memorial tribute", *Physics Today* 16, 10, 58 (1963), 59.
14. In Ajoy Ghatak and S. Lokanthan, *Quantum Mechanics: Theory and Applications*, Dordrecht: Springer 2004, p. 47.
15. Sir James Jeans, *Physics and Philosophy*, New York: Dover Publications 1981 [1943] p. 133.
16. Eric Hobsbawm, *The Age of Extremes: The Short Twentieth Century, 1914–1991*, London: Abacus 1995, p. 539.
17. Werner Heisenberg, *Physics and Philosophy: The Revolution in Modern Science*, New York: Harper & Brothers 1958, pp. 71–72.

18. See n. 9 in this chapter above.
19. In Ivars Peterson, *Newton's Clock: Chaos in the Solar System*, New York: W. H. Freeman & Co 1995, p. 229.
20. George Gamow, *Mr. Tompkins in Paperback*, Cambridge: Cambridge University Press 1993 [1965], p. 65.
21. Steven Weinberg, *Dreams of a Final Theory: The Search for the Fundamental Laws of Nature*, New York: Vintage 1994, p. 77.
22. Petersen, A. "The Philosophy of Niels Bohr", Bulletin of the Atomic Scientists, 19:7 (1963), 12.
23. Marcus Chown, 'Quantum rebel wins over doubters', *New Scientist*, 183(2547), pp. 30–35. Available at: https://www.newscientist.com/article/mg19325915-400-quantum-rebel-wins-over-doubters/ [Accessed on 10 June 2023].
24. William T. Scott, *Erwin Schrödinger: An Introduction to his Writings*, Amherst: University of Massachusetts Press 1967, p. 107.
25. Stephen W. Hawking, *Black Holes and Baby universes and Other Essays*, New York: Bantam Books 1993, p. 45.
26. Karl R. Popper, preface to Franco Selleri, *Le grand débat de la théorie quantique*, Paris: Flammarion 1986, p. 6 (his stress).
27. In the 6th Messenger Lecture, Cornell University 1964. See Richard P. Feynman, *The Character of Physical Law*, Cambridge: MIT Press 1985 [1965], p. 129.
28. Stefanos Trachanas, *Ο κύκλος: επιστήμη και δημοκρατία σε ανήσυχους καιρούς* [*The Circle: Science and Democracy in Uneasy Times*], Heraklion: Crete University Press 2024, p. 362.
29. The Nobel Prize in Physics 2022, https://www.nobelprize.org/prizes/physics/2022/popular-information/ [Accessed on 11 July 2023].

Chapter 5—The Lustrous Dream of Unification

1. Florian Cajori, *Newton's* Principia: *Motte's Translation Revised*. Berkeley: University of California Press 1946, xviii.
2. Stephen Hawking and Roger Penrose, *The Nature of Space and Time*, Princeton: Princeton University Press 1996, p. 26.
3. Edmund O. Wilson, *Consilience: The Unity of Knowledge*, New York: Vintage Books 1998, pp. 4–5.
4. In Tony Hay and Patrick Walters, *The New Quantum universe*, Cambridge: CUP 2003, p. 228.

5. Steven Weinberg, *Dreams of a Final Theory: The Search for the Fundamental Laws of Nature*, New York: Vintage 1994, p. 107.

6. Richard Feynman, *QED: The Strange Theory of Light and Matter*, Princeton: Princeton University Press 1988, p. 10.

7. Richard P. Feynman and Ralph Leighton, *"What do you care what other people think?" Further Adventures of a Curious Character*, New York: W. W. Norton & Company 2001, p. 14.

8. Richard P. Fenyman, *"Surely you're joking Mr. Feynman!" Adventures of a Curious Character*, London: Vintage 1992, p. 21.

9. John Naughton in *The Observer*, as cited on the cover of Richard P. Feynman and Ralph Leighton, *"What do you care what other people think?" Further Adventures of a Curious Character*, New York: W. W. Norton & Company 2001.

10. James Gleick, *Genius: The Life and Science of Richard Feynman*, New York: Open Road 2011 (e-book).

11. Abdus Salam, *Unification of Fundamental Forces: The First of the 1988 Dirac Memorial Lectures*, Cambridge: CUP 1990, p. 35.

12. Stephen Hawking, *A Brief History of Time: From the Big Bang to Black Holes*, Random House 2009 (e-book).

13. Steven Weinberg, *Dreams of a Final Theory: The Search for the Fundamental Laws of Nature*, New York: Vintage 1994, p. 119.

14. John Iliopoulos, personal communication.

15. In Madhusree Mukerjee, "Explaining Everything", *Scientific American*, Vol. 274, No. 1 (January 1, 1996).

Chapter 6—Light Shakes Hands with Life and Art

1. Nikos Kazantzakis *The Odyssey. A Modern Sequel,* trans. Kimon Friar, New York: Simon and Schuster 1958, p. 1.

2. In Michael I. Sobel, *Light*, Chicago: Chicago University Press 1987, p. 168.

3. Leonardo da Vinci, *Notebooks*, ed. Thereza Wells, Oxford: Oxford University Press 2008, p. 105.

4. Alfred North Whitehead, *Science and the Modern World*, Cambridge: Cambridge University Press 2011 [1926], pp. 67–69.

5. Octavio Paz, "Your Eyes", trans. Kevin O'Donnell, http://poetrywit houtborders.blogspot.gr/2009/10/your-eyes.html [Accessed on 12 May 2023].

6. Aeschylus, *Prometheus Bound*, trans. C. Philip Velacott, London: Penguin 2003 (e-book).
7. C. P. Cavafy, "Candles", in *Collected Works*, trans. Edmund Keeley and Philip Sherrard, Princeton: Princeton University Press 1992, p. 13.
8. E. M. Forster, *Alexandria: A History and Guide*, Alexandria 1922, p. 145.
9. From the *New York Herald*, January 1880, in Robert Friedel, Paul Israel and Bernard S. Finn, *Edison's Electric Light: The Art of Invention*, Baltimore: The Johns Hopkins Press 2010, p. 89.
10. Galileo, 'The Sidereal Messenger', in M. A. Finocchiaro (ed.) *The Essential Galileo*. Indianapolis: Hackett Publishing 2008, p. 50.
11. Roland Barthes, *Camera Lucida: Reflections on Photography*, trans. Richard Howard, New York: Hill and Wang 1981, pp. 80–81.
12. Luis Buñuel, *An Unspeakable Betrayal: Selected Writings*, trans. Garrett White, Berkeley: University of California Press 2000, pp. 136–137.
13. Andrei Tarkovsky, *Sculpting in Time: Reflections on the Cinema*, trans. Kitty Hunter-Blair, Austin: University of Texas Press 1987, pp. 187–188.
14. Costanzo Constantini (ed.), *Conversations with Fellini*, trans. Sohrab Sorooshian, Orlando: Harcourt Brace 1995, p. 193.
15. In Allen A. Boraiko and Charles O'Rear, "The laser: A 'splendid light' for man's use", *National Geographic* Vol. 165, No. 3 (March 1984), 336.
16. Sophocles, *Antigone* 332 and 365–366, trans. Reginald Gibbons and Charles Segal, Oxford: Oxford University Press 2003, pp. 68–69.
17. "Γεράσιμος Σκλάβος, ο γλύπτης του φωτός" ["Gerasimos Sklavos, the Sculptor of Light"], *Epta Imeres* special supplement to *Kathimerini* newspaper, 24th January 1999.
18. As above.
19. Yannis Tsarouchis, *Έλληνες ζωγράφοι* [*Greek Painters*], Athens: Kastaniotis 2003, p. 22.
20. Pliny, *Natural History* XXXV, trans. H. Rackham, Loeb Classical Library. Cambridge, MA: Harvard University Press 1961.
21. Odysseas Elytis, *Ανοιχτά χαρτιά* [*Open Papers*], Athens: Ikaros 1996, p. 575.

Chapter 7—The Universe: The Empire of Light

1. Italo Calvino, *Mr. Palomar*, trans. William Weaver, London: Vintage 1999, p. 41.

2. Jorge Luis Borges, "Cosmogony", trans. Jackie Joseph, https://allpoetry.com/poem/15273449-Cosmogonia----translation--by-Jorge-Luis-Borges [Accessed on 21 June 2023].

3. Italo Calvino, "All at One Point" in *The Complete Cosmicomics*, trans. Martin McLaughlin, Tim Parks and William Weaver, New York: Houghton Mifflin 2002, p. 43.

4. Robert Jastrow, *God and the Astronomers*, New York: W. W. Norton 1978, p. 116.

5. Odysseus Elytis, *The Axion Esti*, trans. Edmund Keeley and George Savidis, Pittsburgh: University of Pittsburgh Press 1979, p. 3.

6. NASA's *Cosmic Times* for 1993: "Baby universe's 1st Picture", https://imagine.gsfc.nasa.gov/educators/programs/cosmictimes/online_edition/1993/baby.html [Accessed on 24 July 2023].

7. Antoine de Saint-Exupéry, *The Little Prince*, trans. Katherine Woods, Thorndike: G. K. Hall & Co. 1995, p. 167.

8. Arthur C. Clarke, "The Light of Common Day" in John Carey (ed.), *Eyewitness to Science*, Cambridge, MA: Harvard University Press 1995, pp. 427–435.

9. Richard Feynman, "The Relation of Physics to Other Sciences" in: *The Feynman Lectures on Physics*, Vol. 1, Chapter 3, Caltech online edition at http://www.feynmanlectures.caltech.edu/I_03.html [Accessed on 12 June 2023].

10. Valia Nikolaou and Athina Soutzou, "Σιμόπουλος στον «E.T.» για τις εικόνες του James Webb" ["Simopoulos to *Eleftheros Typos* on the James Webb pictures"], *Eleftheros Typos* newspaper, 13 July 2022.

11. Galileo to Doge Leonardo Donato, 24th August 1609, quoted in J. D. Bernal, *The Social Function of Science*, London: Routledge 1946 [1939], p. 187, n. 2.

12. Leonard Shlain, *Art & Physics: Parallel Visions in Space, Time and Light*, New York: William Morrow 1991, pp. 286–287.

13. Vasilis Xanthopoulos, $Περί αστέρων και συμπάντων$ [*On Stars and universes*], Heraklion: Crete University Press 1985, p. 28.

14. Dionysis Simopoulos, $Κοσμική Οδύσσεια$ [*Cosmic Odyssey*], Athens: Eugenides Foundation Digital Planetarium 2003, p. 24.

15. Kip S. Thorne, *Black Holes and Time Warps: Einstein's Outrageous Legacy*, New York: W. W. Norton & Co. 1995, p. 23.

16. C. P. Cavafy, "Walls", in *Collected Poems,* trans. Edmund Keeley and Philip Sherrard, Princeton: Princeton University Press, 1992, p. 3.

17. Vasiliki Pavlidou, prologue to the Greek edition of Marcia Bartusiak's *Black Hole*, trans. Nestoras Chounos and Panagiotis Drepaniotis, Heraklion: Crete University Press 2020, pp. 16–17.

18. Sakis Ioannidis, "Οι «Φωτογράφοι» της μαύρης τρύπας στην «Κ»" ["Black hole 'photographers' speak to *Kathimerini*"], *Kathimerini* newspaper, 16 May 2022.

19. David Elbaz, *À la recherche de l'univers invisible*, Paris: Odile Jacob 2016, p. 46.

20. Interview with David Weinberg in *Discover* magazine, Vol. 15 (1994), p. 75.

21. Steven Weinberg, *The First Three Minutes: A Modern View of the Origin of the universe*, London: Flamingo 1993 p. 149.

22. Nikos Prantzos, «Το μέλλον του σύμπαντος: μια φευγαλέα αναλαμπή στο σκοτάδι» ["The future of the universe: a fleeting flash in the dark", *Ouranos* magazine, Greek Astronomy and Space Society, issue 100 (August 2016).

Chapter 8—Epilogue: A Farewell to Light

1. Edmund O. Wilson, *Consilience: The Unity of Knowledge*, New York: Vintage Books 1998, p. 8.

Index

A

Aeschylus 6
Aether 38, 60–62, 66, 237
  in modern cosmology 61
Alhazen* 19–21
Al-Khwārizm 18
Al-Kindi* 18–19
Allen, Woody 197
Ampère, André 62
Anaxagoras 27
Anderson, Carl 139
Andromeda
  galaxy 50, 194, 210
  nebula 196
Anthropic principle 219, 240
Antimatter 94, 138–141
Antiphanes 5
Antiquarks 156
Archimedes 17
Aristarchus of Samos 25
Aristotle 11–14, 49
Art and light
  cinema 180–182
  painting* 190–192
  photography 179–182

  sculpture* 187–190
Artificial Intelligence (AI) 231
  and Unified Theory 162
Aspect, Alain 133
Asymptotic freedom* 157–158
Averroes 18
Avicenna 18, 21

B

Bacon, Roger* 23–24
Barish, Barry 104
Barthes, Roland 94, 179
Bartusiak, Marcia 104
Becquerel, Henri* 148
Bell, John 132, 133
  Bell inequalities* 133
Bethe, Hans 217
Big Bang* 198–201
Big Crunch 238, 240, 242
Big Rip 240
Black holes 100, 137, 226,
  228–233, 260
  and quasars 227, 229
  event horizon of 228

in Cygnus* 229
photography of 229, 232
radiation from 232
Sagittarius A* 231
Bohr, Niels
complementarity 120, 129
quantum jumps* 115–116
stability of matter 125
Bohr, Niels* 114–117
Bondi, Hermann 86
Borges, Jorge Luis 89, 198
Born, Max 122
Bouman, Katie 230
Brahe, Tycho 25, 26
Braudel, Fernand 32
Buñuel, Luis 181

C

Calasso, Roberto 5
Calvino, Italo 44, 193, 200
Carroll, Lewis 85
Cavafy, C.P. 172, 227
Celan, Paul 86
Chandrasekhar, Subrahmanyan* 226
Cherenkov radiation 50
Cinema* 180–182
Clarke, Arthur C. 207
Clauser, John 133
Climate crisis 167, 233
COBE satellite 204, 215
Colours 42–45
and quarks 158
in Goethe 47
in Newton 43
in painting 192
in philosophy 48
of stars 208
Coma Berenices* 234
Complementarity 120, 129
Consilience 39
Copenhagen Institute* 116–117
Copenhagen School 128, 133
Copernicus 25, 110

Cosmic radiation (CMB) 3, 62,
201–204, 216, 239
and space missions 205
Cosmological constant 198, 237
Crab nebula* 221–223
Critical density
of the Universe 238
Curie, Marie 148
Curvature of space-time 99
Cygnus 223, 229

D

Dark energy
and future of the Universe 240
and structure of the Universe 239
Dark energy* 235–237
Dark matter 162, 205, 233–235,
237
Darwin, Charles 39
Da Vinci, Leonardo
camera obscura 179
on sight 168
Davisson, C. J. 118
De Broglie, Louis 117, 118–119,
121
Democritus 7, 38, 112, 156
Descartes, René* 30–32
Dicke, Robert 202
Diffraction of light 42
Diffusion of light 41
and colours 45
Dirac, Paul 138–139, 141
Doppler effect 195
Dostoevsky, Fyodor 98

E

Eclipses
lunar 40
solar 40, 101
Eddington, Arthur 101, 226
Einstein, Albert
and EPR experiment 132

and quantum theory* 107–110,
  111–120
biography* 73–76
cosmological constant 197, 237
general theory of relativity 97,
  100–102, 226
photoelectric effect 75, 110, 112
special theory of relativity 76,
  84–86
Elbaz, David 234
Electromagnetic induction 63, 64
Electromagnetic spectrum 171, 207,
  247
Electromagnetic theory 64–66
Electrons
  wave nature of 119, 138
Eliade, Mircea 34
Eliot, T.S. 241
Elytis, Odysseas 46, 128, 191, 203
Empedocles 9, 13
Epicurus 7
Epimenides of Knossos 196
EPR experiment* 132–133
Equivalence principle 98
Euclid 15, 19, 33
Euler, Leonhard 55
European Organization for Nuclear
  Research (CERN) 92, 93, 132,
  153, 160
Event Horizon Telescope (EHT)
  229–232
Expansion of the Universe 197
Extraterrestrial life
  search for 176

F

Faraday, Michael 62–64, 136
Fellini, Federico 181
Fermat, Pierre 32, 33, 41
  Fermat's principle 33
Fermi, Enrico 152
Feynman, Richard 143–146
  Challenger disaster 146

Feynman diagrams 142
  on Maxwell's equations 65
  on quantum mechanics 131
  on the discovery of atoms 113
  Quantum Electrodynamics 141
Field 67
  concept of 63
  electromagnetic 64–66, 67
  gravitational 67, 98–102, 227,
    228
  in Faraday 63
  in Maxwell 64–66
  magnetic 65, 143, 221
  quantum 91, 157, 160
Forster, E.M. 172
Fresnel, Augustin 57
Friedmann, Alexander 197

G

Galaxies
  Andromeda 50, 194, 210
  formation of 204, 205–206, 216
  Milky Way 196, 241
  recession of 193–198, 237
Galileo 27–30, 77, 135, 215
  and light 29
  and telescopes 27, 175
Gamma rays 171, 207, 209
Gamow, George 127, 202
General theory of relativity, see
  Relativity
Gillispie, C. 5
Glashow, Sheldon 154
Gluons 158, 160
Gödel, Kurt 89
God particle, see Higgs boson
Goethe, Johann Wolfgang von 10,
  46–47
Grand Unified Theory 159
Gravitational dilation of time 102
Gravitational waves 103–105, 214,
  235
Gravitons 91

Gravity 147, 149, 159
Grosseteste, Robert* 22–23

H

Hadrons 156
Hawking, Stephen 29, 51, 232–233
    and black holes 229
    and time travel 89
    on quantum physics 137
    on Schrödinger's cat 130
    on the atom bomb 96
    on Unified Theory 159
Heisenberg, Werner 115, 124–125
    and quantum theory 138
Hertz, Heinrich 59
Hesiod
    Theogony 4
Higgs boson* 160–161
Holograms 56, 185, 186
Homer 5, 17, 45
Hooke, Robert 52
    Micrographia 178
Hubble, Edwin 193–197
    Hubble's Law 195, 199
Hubble Space Telescope 177, 211,
    229, 236
Huygens, Christiaan 20, 55–56

I

Iliopoulos, John 157
Inflation theory 205, 239
Interaction
    electromagnetic 147–148, 153,
        155, 158, 160
    gravitational 147, 149, 159
    strong 148–150, 158, 160
    weak 148–150, 152, 155, 160
Interference of light 56

J

Jastrow, Robert 200

Jeans, James 120
Joyce, James
    and quarks 156

K

Kant, Immanuel 193
Kapitas, Pyotr 202
Kazantzakis, Nikos 165
Keats, John 46, 139
Kelvin, Lord* 67–69
Kepler, Johannes 24–27, 223

L

Laplace, Pierre-Simon de 68, 125
Large Hadron Collider (LHC) 160
Large Magellanic Cloud 194, 210,
    223
Lasers 123, 182, 185–187
    stimulated emission 183
Leibniz, Gottfried 52
Lemaître, Georges 200
Length
    contraction of* 90–91
Lenses 174, 176
    in Galileo and Kepler 26
Leptons 156
Leucippus 7
Light
    artificial 171
    primordial 201
    visible 171
LIGO, see Gravitational waves
Lucretius* 8

M

Manichaeism* 17
Mass-energy equivalence 94–95, 101
Maxwell, James Clerk 64–66
    equations 64
Michelson, A. 49, 60, 67

Michelson-Morley experiment
    60–62, 72, 80
    interferometer 60
Microscopes* 178
Milky Way 196, 241
Minkowski, Hermann 82
Morley, E. 60
Muons* 93
Muslim world
    science in* 18–21
Mythology of light* 4–5

N

Neoplatonism 13
NESTOR experiment x
Neutrinos 91, 223, 235
Neutron stars* 220–221
Newton, Sir Isaac 43, 51–55
    and colours 43, 52
Nuclear energy
    atom bomb 96, 227
    fission 95
    fusion 217
    in stars 218, 239

O

Oppenheimer, Robert* 227
Orion nebula 221
Ørsted, Hans Christian
    and electromagnetism 62
Özel, Feryal 231

P

Painting* 190–192
Parmenides* 6
Pascal, Blaise 4, 32
Pauli, Wolfgang 115
Pavlidou, Vasiliki 230
Penzias, Arno 201, 202
Photography* 179–180
Photons 97, 110–112

after the Big Bang 203
and sight 170
entanglement of* 132–133
in Feynman diagrams 142
in laser physics* 183
Photosynthesis* 165–167
Planck, Max* 108–110
Planck's constant 109, 115, 118, 125
Planck space mission 205, 234
Plato 8–11, 124, 210
Pliny the Elder
    Natural History 190
Podolsky, Boris 132
Polarisation of light* 57–58
Pope, Alexander 54
Popper, Karl 130
Population inversion 183
    use in lasers 184
Positrons 97, 139
Prantzos, Nikos 242
Presocratics 5
Principle of least action 33
Psaltis, Dimitris 232
Ptolemy 15–16, 19, 20, 33

Q

Quanta 107–108, 111. *See also*
    Photons
    energy 115
    light 111, 142
Quantum chromodynamics 158
Quantum Electrodynamics (QED)
    141–143, 151, 157
Quantum field, *see* Field, quantum
Quantum leaps 115
Quantum mechanics 33, 69,
    108–109, 121–124, 128, 131,
    138
    and Theory of Everything 52
    and uncertainty principle 124,
    125–129
    applications of 131
    Copenhagen interpretation of 129

entanglement 132–133
problems interpreting* 128–133
Quantum theory
and de Broglie 128
and Einstein 107–110, 111–120, 128
and EPR paradox 132
and Heisenberg 128
and Niels Bohr 114
and Schrödinger 128
of journalism 126
Quantum wave 119
Quarks 156–158
colours 158
flavours 156
Quasars 194
and black holes 227, 229
speed of 196
Quintessence 13, 237

R

Radiation
coherent 183
microwave (CMB) 201–204, 215, 239
Radioactivity 148
Rainbows 10, 43, 46
Rashid, Harun al 18
Red giants 220, 241
Antares 210, 219
Refraction of light 16, 41
rainbows 46
Snell's Law 33
Reize, David 105
Relativity, general theory of 75
and black hole photography 231
and equivalence principle 98
and gravity 75, 98, 136
and time dilation* 102
and time travel 88
confirmation of 101, 112, 231

Relativity, special theory of 61, 68, 72–74, 76, 84–87, 97, 100–102
and length contraction 90
and mass increase 95
and notion of time 73, 82, 246
and speed of light 79
and the aether 77, 80
and time dilation 84, 87, 93
confirmation of 93
first axiom 77, 79
second axiom 78, 79, 82
Resvanis, Leonidas xi
Retina 26, 32, 168–170, 186, 208
Riemann, Bernhard 99
Rømer, Ole 49
Rosen, Nathan 132
Rubbia, Carlo 154
Russell, Bertrand* 109

S

Sagittarius A* 231
Saint-Exupéry, Antoine de 207
Salam, Abdus 149–150, 154
Schrödinger, Erwin 121–122, 124, 138
Schrödinger's cat 130
Sculpture 187–190
Seferis, George 188
Shadows 40
and diffraction 42
Sight 168–171
in Alhazen 20
in al-Kindi 19
in Aristotle 12
in Eculid 15
in Empedocles 6
in Epicurus 7
in Kepler 26
in Leonardo da Vinci 168
in Plato 10
in Ptolemy 16
in St. Augustine 11

in the atomists 14
Simopoulos, Dionysis 212, 225
Simultaneity
    and relativity 83
    concept of 82
Sirius, star 50, 216, 218
Sklavos, Gerasimos* 188–190
Smoot, George 204
Snell, Willebrord 33
Sophocles 6
Special theory of relativity, *see*
    Relativity
Speed of light 48–51
    and relativity 72, 79
    measurement of 49
    reference value 49
Spontaneous symmetry breaking
    155, 160, 161
Stability of matter 114
Standard Model* 159–162
Starlight* 206–209
St. Augustine 49
    On Genesis 11
String theory* 163
St. Thomas Aquinas 13
Sun
    as a red giant 241
    end of 220
    nuclear reactions in 217, 219
Supernovae 222, 225, 236
Szilard, Leo 74

T

Tarkovsky, Andrei 181
Telescopes 26, 175–178
    Arecibo 176
    Atacama (ACT) 199
    EHT 229, 232
    FAST 176
    Hubble 177
    James Webb 212
    LBT 176
    reflector 52, 176

refractor 175
    Skinakas, Crete 213
Theory of Everything 52, 137, 150,
    159, 232
Thomson, J. J. 118
Thorne, Kip 88, 89, 104, 227, 229
Time travel* 86–89
Trachanas, Stefanos 133
Twain, Mark 38
Twin paradox 87

U

Uncertainty (indeterminacy)
    principle 124, 125–128
    black hole 232
Unified Theory 75
Universe
    closed 238, 240
    critical density of 238
    flat 239
    open 239

V

Van der Meer, Simon 154

W

W and Z particles 154
    confirmation of 160
Wave equation 121, 122
Wave nature
    of electrons 118, 138
    of light 37, 42, 44, 58
    of material bodies 119
    of particles 245
Wave-particle duality 112, 119, 120,
    246
Wave phenomena 55
Waves
    electromagnetic 55, 65–66, 71,
        117, 142, 195
    probability 122

quantum 119
Weinberg, Steven 129, 141, 153, 240
Weiss, Rainer 104
Wells, H.G. 87
Wertheim, Margaret 22
Wheeler, John 100, 227
White dwarfs 220
Whitehead, Alfred North 9, 101, 169
Wilson, Edmund O. 39, 137, 248
Wilson, Robert 201, 202
Wittgenstein, Ludwig 48

X
Xanthopoulos, Vasilis 220
X-rays 64, 68, 112, 177, 229

Y
Young, Thomas* 56–57

Z
Zeilinger, Anton 133

Printed in the USA
CPSIA information can be obtained
at www.ICGtesting.com
CBHW071804040824
12688CB00002B/37